ADVANCES IN PLANT PHYSIOLOGY

ADVANCES IN PLANT PHYSIOLOGY

Ramesh Umrani

ANMOL PUBLICATIONS PVT. LTD.
NEW DELHI - 110 002 (INDIA)

ANMOL PUBLICATIONS PVT. LTD.
H.O.: 4374/4B, Ansari Road, Daryaganj,
New Delhi-110 002 (India)
Ph.: 23278000, 23261597
B.O.: No. 1015, Ist Main Road, BSK IIIrd Stage
IIIrd Phase, IIIrd Block,
Bangalore - 560 085 (India)
Visit us at: www.anmolpublications.com

Advances in Plant Physiology

ISBN 978-81-261-3651-3

PRINTED IN INDIA

Printed at Mehra Offset Press, Delhi.

Contents

Preface

The science of plant physiology has been changing rapidly and specialists continue to probe their field in-depth, much of the modern biological science is intimately intertwined with the rest of the science. The profound impact of molecular genetics is clearly evident in plant physiology as well as the whole of biology. The book "Advances in Plant Physiology" takes into account the vast increase in physiological knowledge and the evolution of fundamental concepts that are uniquely associated with plant physiology.

To help students understand the scientific approach and develop an attitude of critical understanding, some of the fundamental laboratory terms are included. Each chapter is vividly illustrated to make it more understandable to students, research and extension workers, planners, administrators and other end users citing pertinent references.

Chapter 1

Introduction

Plant physiology is a subdiscipline of botany concerned with the function, or physiology, of plants. Closely related fields include plant morphology (structure of plants), plant ecology (interactions with the environment), phytochemistry (biochemistry of plants), cell biology and molecular biology.

Fundamental processes such as photosynthesis, respiration, plant nutrition, plant hormone functions, tropisms, nastic movements, photoperiodism, photo-morphogenesis, circadian rhythms, environmental stress physiology, seed germination, dormancy and stomata function and transpiration, both part of plant water relations, are studied by plant physiologists.

SCOPE

The field of plant physiology includes the study of all the internal activities of plants—those chemical and physical processes associated with life as they occur in plants. This includes study at many levels of scale of size and time. At the smallest scale are molecular interactions of photosynthesis and internal diffusion of water, minerals, and nutrients. At the largest scale are the processes of plant development, seasonality, dormancy, and reproductive control.

Major subdisciplines of plant physiology include phytochemistry (the study of the biochemistry of plants) and phytopathology (the study of disease in plants). The scope of plant physiology as a discipline may be divided into several major areas of research.

First, the study of phytochemistry (plant chemistry) is included within the domain of plant physiology. In order to function and survive, plants produce a wide array of chemical compounds not found in other organisms. Photosynthesis requires a large array of pigments, enzymes, and other compounds to function.

Because they cannot move, plants must also defend themselves chemically from herbivores, pathogens and competition from other plants. They do this by producing toxins and foul-tasting or smelling chemicals. Other compounds defend plants against disease, permit survival during drought, and prepare plants for dormancy. While other compounds are used to attract pollinators or herbivores to spread ripe seeds.

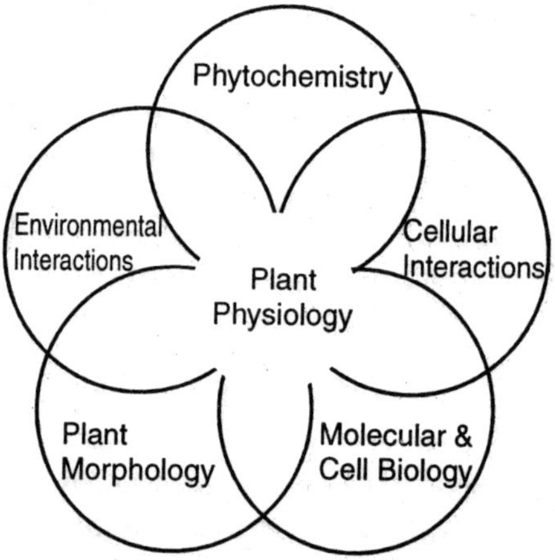

Fig. Key areas of plant physiology

Secondly, plant physiology includes the study of biological and chemical processes of individual plant cells. Plant cells have a number of features that distinguish them from cells of animals, and which lead to major differences in the way that plant life behaves and responds differently from animal life. For example, plant cells have a cell wall which restricts the shape of plant cells and thereby limits the flexibility and mobility of plants. Plant cells also contain chlorophyll, a chemical compound that interacts with light in a way that enables plants to manufacture their own nutrients rather than consuming other living things as animals do.

Thirdly, plant physiology deals with interactions between cells, tissues, and organs within a plant. Different cells and tissues are physically and chemically specialized to perform different functions. Roots and rhizoids function to anchor the plant and acquire minerals in the soil.

Leaves function to catch light in order to manufacture nutrients. For both of these organs to remain living, the minerals acquired by the roots must be transported to the leaves and the nutrients manufactured in the leaves must be transported to the roots. Plants have developed a number of means by which this transport may occur, such as vascular tissue, and the functioning of the various modes of transport is studied by plant physiologists.

Fourthly, plant physiologists study the ways that plants control or regulate internal functions. Like animals, plants produce chemicals called hormones which are produced in one part of the plant to signal cells in another part of the plant to respond. Many flowering plants bloom at the appropriate time because of light-sensitive compounds that respond to the length of the night, a phenomenon known as photoperiodism. The ripening of fruit and loss of leaves in the winter are controlled in part by the production of the gas ethylene by the plant.

Finally, plant physiology includes the study of how plants respond to conditions and variation in the environment, a field known as environmental physiology. Stress from water loss, changes in air chemistry, or crowding by other plants can lead to changes in the way a plant functions. These changes may be affected by genetic, chemical, and physical factors.

BIOCHEMISTRY OF PLANTS

are primarily constructed—carbon, oxygen, hydrogen, calcium, phosphorus, etc.—is not different from similar lists for animals, fungi, or even bacteria. The fundamental atomic components of plants are the same as for all life; only the details of the way in which they are assembled differs.

Despite this underlying similarity, plants produce a vast array of chemical compounds with unusual properties which they use to cope with their environment. Pigments are used by plants to absorb or detect light, and are extracted by humans for use in dyes. Other plant products may be used for the manufacture of commercially important rubber or biofuel. Perhaps the most celebrated compounds from plants are those with pharmacological activity, such as salicylic acid (aspirin), morphine, and digitalis. Drug companies spend billions of dollars each year researching plant compounds for potential medicinal benefits.

CONSTITUENT ELEMENTS

Plants require some nutrients, such as carbon and nitrogen, in large quantities to survive. Such nutrients are termed macronutrients, where the prefix *macro-* (large) refers to the quantity needed, not the size of the nutrient particles themselves. Other nutrients, called micronutrients, are required only in trace amounts for plants to remain healthy. Such micronutrients are usually absorbed as ions dissolved in water taken from the soil, though carnivorous plants acquire some of their micronutrients from captured prey.

The following tables list element nutrients essential to plants. Uses within plants are generalized.

Tabel: Macronutrients (Necessary in large quantities)

Element	Form of uptake	Notes
Nitrogen	NO_3^-, NH_4^+	Nucleic acids, proteins, hormones, etc.
Oxygen	O_2 H_2O	Various organic compounds
Carbon	CO_2	Various organic compounds
Hydrogen	H_2O	Various organic compounds
Potassium	K^+	Cofactor in protein synthesis, water balance, etc.
Calcium	Ca^{2+}	Membrane synthesis and stabilization
Magnesium	Mg^{2+}	Element essential for chlorophyll

Phosphorus	$H_2PO_4^-$	Nucleic acids, phospholipids, ATP
Sulfur	SO_4^{2-}	Constituent of proteins and coenzymes

Micronutrients. (Necessary in small quantities)

Element	Form of uptake	Notes
Chlorine	Cl^-	Photosystem II and stomata function
Iron	Fe^{2+}, Fe^{3+}	Chorophyll formation
Boron	HBO_3	Crosslinking pectin
Manganese	Mn^{2+}	Activity of some enzymes
Zinc	Zn^{2+}	Involved in the synthesis of enzymes and chlorophyll
Copper	Cu^+	Enzymes for lignin synthesis
Molybdenum	MoO_4^{2-}	Nitrogen fixation, reduction of nitrates
Nickel	Ni^{2+}	Enzymatic cofactor in the metabolism of nitrogen compunds

PIGMENTS

Among the most important molecules for plant function are the pigments. Plant pigments include a variety of different kinds of molecules, including porphyrins, carotenoids, and anthocyanins. All biological pigments selectively absorb certain wavelengths of light while reflecting others. The light that is absorbed may be used by the plant to power chemical reactions, while the reflected wavelengths of light determine the colour the pigment will appear to the eye.

Chlorophyll is the primary pigment in plants; it is a porphyrin that absorbs red and blue wavelengths of light while reflecting green. It is the presence and relative abundance of chlorophyll that gives plants their green colour. All land plants and green algae possess two forms of this pigment: chlorophyll *a* and chlorophyll *b*. Kelps, diatoms, and other photosynthetic heterokonts contain chlorophyll *c* instead of *b*, while red algae possess only chlorophyll *a*. All chlorophylls serve as the primary means plants use to intercept light in order to fuel photosynthesis.

Carotenoids are red, orange, or yellow tetraterpenoids. They function as accessory pigments in plants, helping to fuel photosynthesis by gathering wavelengths of light not readily absorbed by chlorophyll. The most familiar carotenoids are carotene (an orange pigment found in carrots), lutein (a yellow pigment found in fruits and vegetables), and lycopene (the red pigment responsible for the colour of tomatoes). Carotenoids have been shown to act as antioxidants and to promote healthy eyesight in humans.

Anthocyanins (literally "flower blue") are water-soluble flavonoid pigments that appear red to blue, according to pH. They occur in all tissues of higher plants, providing colour in leaves, stems, roots, flowers, and fruits, though not always in sufficient quantities to be noticeable. Anthocyanins are most visible in the petals of flowers, where they may make up as much as 30% of the dry weight of the tissue.

They are also responsible for the purple colour seen on the underside of tropical shade plants such as *Tradescantia zebrina*; in these plants, the anthocyanin catches light that has passed through the leaf and reflects it back towards regions bearing chlorophyll, in order to maximize the use of available light.

Betalains are red or yellow pigments. Like anthocyanins they are water-soluble, but unlike anthocyanins they are indole-derived compounds synthesized from tyrosine. This class of pigments is found only in the Caryophyllales (including cactus and amaranth), and never co-occur in plants with anthocyanins.

Betalains are responsible for the deep red colour of beets, and are used commercially as food-colouring agents. Plant physiologists are uncertain of the function that betalains have in plants which possess them, but there is some preliminary evidence that they may have fungicidal properties.

SIGNALS AND REGULATORS

Plants produce hormones and other growth regulators which act to signal a physiological response in their tissues. They also produce compounds such as phytochrome that are sensitive to light and which serve to trigger growth or development in response to environmental signals.

PLANT HORMONES

Plant hormones, also known as plant growth regulators (PGRs) or phytohormones, are chemicals that regulate a plant's growth. According to a standard animal definition, hormones are signal molecules produced at specific locations, that occur in very low concentrations, and cause altered processes in target cells at other locations. Unlike animals, plants lack specific hormone-producing tissues or organs. Plant hormones are often not transported to other parts of the plant and production is not limited to specific locations.

Plant hormones are chemicals that in small amounts promote and influence the growth, development and differentiation of cells and tissues. Hormones are vital to plant growth; effecting processes in plants from flowering to seed development, dormancy, and germination. They regulate which tissues grow upwards and which grow downwards, leaf formation and stem growth, fruit development and ripening, as well as leaf abscission and even plant death.

The most important plant hormones are abscissic acid (ABA), auxins, gibberellins, and cytokinins, though there are many other substances that serve to regulate plant physiology.

PHOTOMORPHOGENESIS

While most people know that light is important for photosynthesis in plants, few realise that plant sensitivity to light plays a role in the control of plant structural development (morphogenesis). The use of light to control structural development is called photomorphogenesis, and is dependent upon the presence of specialized photoreceptors, which are chemical pigments capable of absorbing specific wavelengths of light.

Plants use four kinds of photoreceptors: phytochrome, cryptochrome, a UV-B photoreceptor, and protochlorophyllide *a*. The first two of these, phytochrome and cryptochrome, are photoreceptor proteins, complex molecular structures formed by joining a protein with a light-sensitive pigment. Cryptochrome is also known as the UV-A photoreceptor, because it absorbs ultraviolet light in the long wave "A" region. The UV-B receptor is one or more compounds that have yet to be identified with certainty, though some evidence suggests carotene or riboflavin as candidates. Protochlorophyllide *a*, as its name suggests, is a chemical precursor of chlorophyll.

The most studied of the photoreceptors in plants is phytochrome. It is sensitive to light in the red and far-red region of the visible spectrum. Many flowering plants use it to regulate the time of flowering based on the length of day and night (photoperiodism) and to set circadian rhythms. It also regulates other responses including the germination of seeds, elongation of seedlings, the size, shape and number of leaves, the synthesis of chlorophyll, and the straightening of the epicotyl or hypocotyl hook of dicot seedlings.

Many flowering plants use the pigment phytochrome to sense seasonal changes in day length, which they take as signals to flower. This sensitivity to day length is termed photoperiodism. Broadly speaking, flowering plants can be classified as long day plants, short day plants, or day neutral plants, depending on thir particular response to changes in day length. Long day plants require a certain minimum length of daylight to initiate flowering, so these plants flower in the spring or summer.

Conversely, short day plants will flower when the length of daylight falls below a certain critical level. Day neutral plants do not initiate flowering based on photoperiodism, though some may use temperature sensitivity (vernalization) instead.

Although a short day plant cannot flower during the long days of summer, it is not actually the period of light exposure that limits flowering. Rather, a short day plant requires a minimal length of uninterrupted darkness in each 24 hour period (a short daylength) before floral development can begin.

It has been determined experimentally that a short day plant (long night) will not flower if a flash of phytochrome activiting light is used on the plant during the night.

Plants make use of the phytochrome system to sense day length or photoperiod. This fact is utilized by florists and greenhouse gardeners to control and even induce flowering out of season, such as the *Poinsettia*.

ENVIRONMENTAL PHYSIOLOGY

Paradoxically, the subdiscipline of environmental physiology is on the one hand a recent field of study in plant ecology and on the other hand one of the oldest. Environmental phyiology is the preferred name of the subdiscipline among plant physiologists, but it goes by a number of other names in the applied sciences. It is roughly synonymous with ecophysiology, crop ecology, horticulture, and agronomy. The particular name applied to the subdiscipline is specific to the viewpoint and goals of research. Whatever name is applied, it deals with the ways in which plants respond to their environment and so overlaps with the field of ecology.

Environmental physiologists examine plant response to physical factors such as radiation (including light and ultraviolet radiation), temperature, fire, and wind. Of particular importance are water relations and the stress of drought or inundation, exchange of gases with the atmosphere, as well as the cycling of nutrients such as nitrogen and carbon.

Environmental physiologists also examine plant response to biological factors. This includes not only negative interactions, such as competition, herbivory, disease, and parasitism, but also positive interactions, such as mutualism and pollination.

TROPISMS AND NASTIC MOVEMENTS

Plants may respond both to directional and nondirectional stimuli. A response to a directional stimulus, such as gravity or sunlight, is called a tropism. A response to a nondirectional stimulus, such as temperature or humidity, is a nastic movement.

Tropisms in plants are the result of differential cell growth, in which the cells on one side of the plant elongate more than those on the other side, causing the part to bend toward the side with less growth. Among the common tropisms seen in plants is phototropism, the bending of the plant toward a source of light. Phototropism allows the plant to maximize light exposure in plants which require additional light for photosynthesis, or to minimize it in plants subjected to intense light and heat. Geotropism allows the roots of a plant to determine the direction of gravity and grow downwards. Tropisms generally result from an interaction between the environment and production of one or more plant hormones.

In contrast to tropisms, nastic movements result from changes in turgor pressure within plant tissues, and may occur rapidly. A familiar example is thigmonasty (response to touch) in the Venus fly trap, a carnivorous plant. The traps consist of modified leaf blades which bear sensitive trigger hairs. When the hairs are touched by an insect or other animal, the leaf folds shut.

This mechanism allows the plant to trap and digest small insects for additional nutrients. Although the trap is rapidly shut by changes in internal cell pressures, the leaf must grow slowly in order to reset for a second opportunity to trap insects.

PLANT DISEASE

Economically, one of the most important areas of research in environmental physiology is that of phytopathology, the study of diseases in plants and the manner in which plants resist or cope with infection. Plant are susceptible to the same kinds of disease organisms as animals, including viruses, bacteria, and fungi, as well as physical invasion by insects and roundworms.

Because the biology of plants differs from animals, their symptoms and responses are quite different. In some cases, a plant can simply shed infected leaves or flowers to prevent to spread of disease, in a process called abscission. Most animals do not have this option as a means of controlling disease. Plant diseases organisms themselves also differ from those causing disease in animals because plants cannot usually spread infection through casual physical contact. Plant pathogens tend to spread via spores or are carried by animal vectors.

One of the most important advances in the control of plant disease was the discovery of Bordeaux mixture in the nineteenth century. The mixture is the first known fungicide and is a combination of copper sulfate and lime. Application of the mixture served to inhibit the growth of downy mildew that threatened to seriously damage the French wine industry.

HISTORY

Sir Francis Bacon published one of the first plant physiology experiments in 1627 in the book, *Sylva Sylvarum.* Bacon grew several terrestrial plants, including a rose, in water and concluded that soil was only needed to keep the plant upright. Jan Baptist van Helmont published what is considered the first quantitative experiment in plant physiology in 1648.

He grew a willow tree for five years in a pot containing 200 pounds of oven-dry soil. The soil lost just two ounces of dry weight and van Helmont concluded that plants get all their weight from water, not soil. In 1699, John Woodward published experiments on growth of spearmint in different sources of water. He found that plants grew much better in water with soil added than in distilled water.

Stephen Hales is considered the Father of Plant Physiology for the many experiments in the 1727 book; though Julius von Sachs unified the pieces of plant physiology and put them together as a discipline. His *Lehrbuch der Botanik* was the plant physiology bible of its time.

Researchers discovered in the 1800s that plants absorb essential mineral nutrients as inorganic ions in water. In natural conditions, soil acts as a mineral nutrient reservoir but the soil itself is not essential to plant growth. When the mineral nutrients in the soil are dissolved in water, plant roots absorb nutrients readily, soil is no longer required for the plant to thrive. This observation is the basis for hydroponics, the growing of plants in a water solution rather than soil, which has become a standard technique in biological research, teaching lab exercises, crop production and as a hobby.

CURRENT RESEARCH

One of the leading journals in the field is *Plant Physiology*, started in 1926. All its back issues are available online for free. Many other journals often carry plant physiology articles, including *Physiologia Plantarum, Journal of Experimental Botany, American Journal of Botany, Annals of Botany, Journal of Plant Nutrition* and *Proceedings of the National Academy of Sciences.*

In horticulture and agriculture along with food science, plant physiology is an important topic relating to fruits, vegetables, and other consumable parts of plants. Topics studied include: *climatic* requirements, fruit drop, nutrition, ripening, fruit set. The production of food crops also hinges on the study of plant physiology covering such topics as Optimal planting and harvesting times and post harvest storage of plant products for human consumption and the production of secondary products like drugs and cosmetics.

Chapter 2

Shoot Part

IT has been already noticed that the seedling bean plant consists of a descending portion, the root, and an ascending part which comes above ground. The latter is known as the primary shoot, and consists of an axis—the stem—upon which are arranged a series of lateral appendages which are called leaves. The points on the stem to which the leaves are attached are usually slightly thickened, and are called nodes or 'joints' the lengths of stem between them being termed internodes.

Flowers ultimately arise upon the shoot, and it is one of the characteristics of seed-bearing plants that seeds are always produced on their shoots and never on roots. For the present, however, flowers may be left for future consideration, and attention paid to the origin and nature of the vegetative shoot or stem with its ordinary green leaves.

In the earliest stages of the development of a bean plant the primary shoot is very short and bears the cotyledons or primary leaves, its tip ending in the plumule. The latter is a bud, and at the time when the seed commences to germinate, its parts cannot be fully made out by observations with the naked eye. As soon as it comes above ground, however, the bud is found, on examination, to consist of a short stem hidden by a number of enfolding leaves.

As growth proceeds, the short stem inside the bud elongates, and the leaves, which at first are crowded upon it, become separated from each other. Marks made upon the stem as previously explained for the root, reveal the fact that the increase in length takes place at the tip of the shoot.

After reaching a certain length the lowest intervals between the leaves cease to elongate; the upper, younger and shorter ones also lengthen and cease in a similar manner, to be followed in turn by still younger parts of the stem nearer the tip. The stem, before the growing season is over, may thus reach a height of two or three feet, or even more, the extreme tip, or growing-point as it is called, remaining young all the time, and acting as a manufactory for the addition of more stem and leaves.

The growing-point, which is of a tender and delicate nature, is protected by the enfolding young leaves, the latter arising as outgrowths from its external surface. The youngest leaves are always nearest the tip the stem which bears them, the older ones being further removed from it in regular order—that is, they arise in acropetal succession, and adventitious leaves are never met with.

Instead of the latter growing at once into a long shoot, bearing leaves at some distance from each other, the primary axis within the plumule elongates very little, its internodes remain very short, and the leaves arising upon it appear crowded together, usually in the form of a rosette, a short distance above where the cotyledons were placed; this form of stem with short contracted inter-nodes is well illustrated in the first season's growth of mangels, turnips, carrots, certain thistles, and red clover.

In such plants as these, the primary root and hypocotyl become much thickened by the deposition within them of reserve-food prepared by the leaves, and it is only during the following year that the growing-point of the stem, which is hidden in the centre of the rosette, elongates and produces a shoot with long internodes, and bearing a series of new leaves at considerable intervals.

In the onion and many bulbous plants the primary stem also remains very short, and the reserve-foods prepared by them are deposited in the bases of the leaves, instead of being stored in the root and stem, as in the former instances.

BUDS

The stems and leaves of all flowering plants originate from buds in the manner indicated above; buds may therefore be termed embryonic or incipient shoots. It is by their growth that trees, which appear so bare in winter, become clothed with fresh green leaves in the succeeding spring. The relationship which they bear to the leaves and stems produced by them is easily discerned by examining the structure and watching the development of the terminal bud of a young sycamore tree.

On the outside is observed a series of scaly leaves, which overlap each other, and protect and cover up the delicate growing point of the twig. A section through the bud shows the disposition of these scaly leaves and within are also seen the ordinary leaves arranged upon a very short stem (s). In spring the inner scaly leaves grow for a time. The stem which bears the rudimentary green foliage-leaves, elongates, and the latter are pushed out from between the protective scaly leaves of the bud . After a week or ten days, the stem has reached a considerable length, and the leaves, which were rudimentary and packed away in the bud, unfold themselves and grow out flat.

The number of foliage-leaves present upon a developed shoot is often indicated in the bud, but in some plants, especially those of a herbaceous character, the growing point of the bud continues to produce new leaves until frost checks it in autumn.

The vegetative shoots of plants usually end in terminal buds and an examination of almost any kind of plant shows that not only are buds present at the tips of the stems, but on their sides as well. These lateral buds arise ordinarily in the upper angles formed where the leaf-bases and stem join each other. The angles are termed the axils of the leases, and the buds are designated axillary buds. Most frequently only one bud is produced in each leaf-axil, but in some instances, two or more may be present.

Generally the first leaves of the bud which are outermost or lowest down on the stem, are rudimentary structures, smaller and different in appearance from those which unfold later. In the primary bud or plumule of the bean, and many similar herbaceous plants, this is observable, but it is most evident in buds which are met with upon perennials, such as shrubs and trees.

In the latter the outermost leaves of the buds are generally more or less firm, tough structures, termed scales or scale-leaves, which protect the interior of the bud from being injured by frost, rain, and other agents during the winter. Buds, such as those of the sycamore and pear, having scales are termed scaly buds, those without, such as mealy guelder-rose, being known as naked buds.

Buds similar to those of the bean and sycamore, previously described, which develop into shoots bearing green foliage-leaves, are termed leaf-buds: when met with upon trees they are sometimes named wood-buds, as it is from them that new woody twigs are produced. Many buds, however, on opening, give rise to flowers only, and are termed flower-buds: a third kind is met with producing short shoots bearing both green leaves and flowers; these are mixed-buds. Among gardeners the two latter forms are known as fruit-buds as it is from them that fruit is obtained.

It is not possible in all cases to distinguish fruit-buds and wood-buds by their outward appearance, although for budding and pruning operations and the general management of fruit-trees it is desirable to do so. In apples and pears the wood-buds are small and pointed, the fruit-buds being blunter, more plump, and of larger size. In cherries and plums both kinds are very similar in appearance, and it is only in spring when they begin to develop that the stouter and blunter characters of the fruit-buds show themselves.

Their position upon the shoots is a great aid in distinguishing the two classes of buds.

BRANCHING OF STEMS

The axis or stem of the primary shoot of plant is at first single, and may continue to grow as a simple straight structure. Usually, however, after a time, branches or secondary axes arise upon it, and these in all cases proceed from buds. The primary bud of a bean, secondary lateral buds are seen in the axils of the leaves of & and b: these are flower-buds, and consequently do not produce long leafy shoots; but secondary axes bearing green leaves frequently occur in the bean, and are generally produced from buds in the axils of the cotyledons.

In many plants the buds in the axils of each leaf of the primary stem develop into leafy shoots, and upon the latter branches may again arise in a similar manner. The total number of stems bearing leaves may thus become very large on a single plant. In the best fodder crops, where large yield is always a desirable feature, branching is exhibited in high degree, and the same may be observed in trees of all kinds, and many weeds, such as groundsel and chickweed.

The main stem of a plant is spoken of as a primary axis, or axis of the first order, the branches upon it being secondary axes, or axes of the second order, those borne by the latter tertiary axes, and so on. For purposes of convenience in description, any axis may be considered a main one, its branches then being secondary axes.

When a stem continues to grow at its apex, for a long time it is spoken of as indefinite in growth: the branches upon it are usually many in number, and smaller than the main stem. This form of branching is spoken of as racemose.

In many plants the terminal bud produces a flower or a collection of flowers, and the main axis such a stem is definite in growth. When lateral branches arise upon it, they are generally few in number, and soon equal or exceed the main stem in vigour. Branching of stems of definite growth is said to be cymose it often resembles. Cymose branching, however, sometimes leads to the formation of what at first sight appears to be a simple main axis of indefinite growth, but which is in reality composed of a series of short axes of different orders.

The amount of growth of twigs during one year or one growing-season is represented by the length between two sets of bud-scale scars.

As the scars are often visible upon the bark for several years they are useful aids in the determination of the age of any length of tree, stem, or twig. Frequently small buds are present in the axils of bud-scales, and as the internodes between the latter remain short such buds appear crowded together upon the twigs and are often visible after the scars have been obliterated.

The length of stem which a bud produces during a year's growth is very varied, some leaf-buds giving rise to shoots not more than a small fraction of an inch long, while others reach a length of several feet. Much depends upon the kind of plant, its age, treatment, and the position of the bud upon the tree, as well as upon external circumstances, such as climate and soil In trees which are unmolested the length of the shoots produced each year by the terminal buds goes on increasing from extreme youth onwards until a certain age is reached, after which the yearly length of the shoots begins to diminish.

The age at which the growth is at a maximum is different for different trees, some forming their longest shoots when they are 15 to 20 years old, others not until 30 to 40 years have elapsed. In old age the large number of buds to be supplied with water and food-constituents, and their increasing distance from the water-supply in the ground, prevents the extensive growth which is noticed in youth: the shoots upon aged trees are therefore very short.

The difference in the general appearance between young and old trees is striking; so long as long shoots are produced the crown or head remains open and largely composed of long straight branches, but when the formation of short shoots begins the crown assumes a denser aspect.

In most trees the terminal bud of a shoot usually develops the strongest shoot, the lateral buds giving rise to shorter branches in regular decreasing order from the tip to the base, where the buds usually produce very short shoots or none at all. In the ash and willow, however, the branches on a shoot are much the same size from top to bottom, and in a few instances the branches are short near the tip and base and long near the middle of the shoot In good soil and a favourable climate the branches of trees are longer than where the ground is poor and lacking in moisture or where the climate is severe.

Nitrogenous manures and absence of light due to overcrowding tend to the production of long shoots, while the bearing of fruit checks the vigour of trees and leads to the formation of short shoots.

Three fairly distinct classes of trees may, however, be recognised. Some trees, such as the peach, bear almost entirely on long shoots one year old and have few or no spurs, while others produce their fruit-buds chiefly at the apex or on the sides of spurs, the long shoots of the tree bearing only wood-buds in their first year; a third group bears almost equally both upon long shoots and spurs.

The apple and pear produce fruit-buds chiefly upon spurs, and rarely upon long shoots which are only one season old. A few varieties of apples, however, such as Cox's Orange pippin, Ribston pippin, and Irish peach, sometimes produce fruit-buds freely on the long shoots of last season.

The plum bears largely upon spurs, but sometimes the fruit-buds may appear upon its young long shoots: when the latter happens they are usually accompanied by wood-buds placed on each side.

The red currant carries its fruit chiefly upon spurs; the black currant, both on young long shoots and spurs, but chiefly on the former.

In the black and white Hearts, Bigarreau and Duke cherries, the fruit-buds are mostly met with upon spurs, but the Morello and Kentish types bear largely upon the one-year-old long shoots.

The gooseberry and apricot resemble the black currant in the arrangement of their fruit-buds.

The raspberry bears upon leafy shoots, which arise in summer from buds on the previous year's cane. The canes or stems which come above ground are biennial. The fruiting cane dies down in autumn, but before this takes place the buds at its base on the underground rootstock grow up into canes: in the following year the buds upon the latter open out into leafy shoots which bear the fruit, after which these canes die away, and are followed by a new set of young canes which originate in a similar manner.

There is considerable difference in the age to which fruit-spurs attain j some, like those of apple and pear, live many years; upon red currant they remain productive longer than upon the black currant; the spurs on the Morello cherry are shorter lived than those upon the other varieties. In pruning those trees with spurs of short life, endeavour should be made to secure relays of young long shoots at frequent intervals.

DORMANT BUDS

On examining trees in spring when the buds are beginning to develop, it will be observed that some of them remain inactive and continue in this condition all the summer. Not only may they refuse to grow in what may be termed their proper season, but they frequently remain undeveloped for long periods. Such buds are termed dormant or resting buds, and are met with upon almost all kinds of plants, chiefly near the base of the stems.

Although many dormant buds soon die, some remain capable of development for several years after their formation, and may give rise to what are termed deferred shoots. In fruit-trees they are termed c water-sprouts' j if they spring from beneath the surface of the ground they are known as ' suckers.' They not uncommonly arise upon 'stocks' which have been grafted or budded.

Destruction of the terminal and lateral buds near the top of a stem tends to promote the growth of deferred shoots from dormant buds at its base. Moreover, pinching out the terminal buds of herbaceous and other plants is often practised with a view to insure the development of all the lateral buds upon it, and the formation of a bushy plant instead of one with a single main stem and few branches.

The grazing and mowing of grasses promotes the full development of all their buds, and a consequent increase of leafy stems. Not only does cutting away or pinching flff the terminal buds promote the development of basal buds likely to become dormant, but anything which impedes the movement of water or 'flow of sap' to the terminal and highly-placed buds tends towards the same result.

In the early formation of cordon fruit-trees, where it is important that all the buds upon the main stem should develop shoots or short spurs, bending the shoot for a time is practised in order to promote the ' breaking' of those buds at the base of the stem which might otherwise remain dormant and leave a length of unfruitful wood.

Adventitious Buds

Dormant buds, are buds which have arisen in regular order in the axils of leaves, but which have remained inactive some time; the only irregularity about them is their period of development.

Buds may, however, arise at any point of a plant, not necessarily in the axil of a leaf, but on any part of the stem, or even upon roots and leaves: such are termed adventitious buds. Examples are met with on the roots of docks, poplars, roses, and many other plants, especially when the upper bud-bearing parts have been removed. They frequently arise and produce shoots upon stems which have been injured. In some instances they proceed from the callus covering the wounds where branches have been cut off; some of the shoots of ' pollard ' trees spring from adventitious buds originating in this manner.

Adventitious buds are often produced upon leaves which have been removed from the parent and pegged down on moist sand or loam. Gardeners take advantage of this peculiarity in propagating begonias.

Similar buds occur upon some kinds of leaves when they are severed from the plant and their petioles stuck in moist ground: the scales of the hyacinth and other bulbs give rise to new plants in this manner.

STEMS AND THEIR VARIETIES

Stems which are soft and which usually last but a short time, are termed herbaceous practically all our annuals have stems of this nature, and many perennial plants also, e.g. nettles and hops. Most stems which last several seasons develop within themselves considerable quantities of wood, and are harder and firmer in consequence: such stems are said to be woody. It must be pointed out, however, that herbaceous stems in reality also possess wood, but only in the form of thin strands, which are relatively small in amount when compared with the remaining soft parts.

All stems, moreover, are soft and herbaceous when very young, so that no real distinction exists between herbaceous and woody stems, as it is a matter of degree of development of the wood within them: a wall-flower or a rose, may be soft and herbaceous in its upper parts and hard and woody below.

Trees and Shrubs have well-developed woody stems, the former possessing a single main stem or trunk, which is devoid of branches for some distance above the ground; the latter have no very distinct main stem, and the chief branches are all much the same in thickness and spring from a point either on or close to the ground.

Many plants have stems which are too weak to maintain an erect position; they consequently grow along the surface of the soil. Some plants have weak stems which always remain prostrate^ while others, designated climbing plants^ have stems which, although too weak to stand upright of themselves, are nevertheless able to use suitable objects near them as supports. Climbing plants support themselves in various ways. In ivy, adventitious roots are developed on one side of the stem, and these serve to fix the plant to bark of trees, walls, and rocks.

Tropaeolums of gardens and wild clematis are supported by their leaves, the petioles of which curve round the stronger branches of plants growing near them. Peas and vetches are also enabled to climb by means of their leaves, some of the leaflets of which are modified into thin threadlike structures termed tendrils. The latter are sensitive to contact, and wind round any slender object which they touch. Plants such as the blackberry, rose, are supported by means of their stiff prickles.

In twining plants the whole stem upholds itself by twisting round neighbouring objects. The stems of some of them always twine to the right when growing round a support; the hop is an example: others, such as bindweed, twine to the left.

A number of peculiar modifications of shoots are met with,many of which receive special names:

In the wild pear, wild plum, hawthorn, sloe, and buckthorn, some of the branches end in hard, sharp points, termed thorns or spines. That they are modified shoots is seen from the fact that they arise in the axils of leaves, and also themselves bear leaves and lateral buds in some instances.

A runner or stolon is a shoot which extends horizontally over the surface of the ground. Its internodes are long, and from its nodes adventitious roots are produced, and grow into the soil. The buds present on the runner then become fixed to the ground, and, developing into upright shoots, form separate plants as soon as the internodes at s die away or are severed.

Strawberry runners and those of creeping crowfoot are good examples. Stems within the soil sometimes resemble roots, but they can be distinguished from the latter by the possession of leaves and buds, and by their originating in the axils of leaves.

A rhizome or ' rootstock ' is an underground shoot, which grows more or less horizontally. Adventitious roots arise at the nodes, and the internodes may be long or short, thick or thin.

Most rhizomes are, however, definite in growth, the main axis, after growing a longer or shorter distance below, comes above ground, the continuation of the rhizome within the soil being carried on by lateral branches. In perennial rhizomes of definite growth, such as those of sedges, grasses, and many other plants, the permanent part which remains below ground is a false main-axis or sym-podium.

The term sucker is applied to any adventitious shoot which originates below ground on the stems or roots of shrubs and trees. It possesses adventitious roots and by separation from the parent may become a new individual plant. Suckers often develop very rapidly and rob the parent of water and nutriment, so that except for purposes of propagation they should be destroyed.

A tuber is a shoot with a short, thick, fleshy stem, and minute scaly leaves, in whose axils are buds or 'eyes.' The most common tubers are developed below ground—e.g. those of potato and Jerusalem artichoke—but they may occur on parts of plants above the soil. The scaly leaves are not visible on the fully-developed potato tuber, as they drop off or shrivel up before ripening is accomplished.

A corm is a short, thick, fleshy stem, with a few thin, scaly leaves covering it, and bearing one or more buds at its apex. Examples are seen in the ordinary crocus and gladiolus of gardens.

The terminal bud has grown in spring into a short stem, bearing on its sides thin, membranous leaves and ordinary green foliage leaves which come above ground. One or more flowers are produced from the axils of the leaves. The substances stored in the stem of the corm are used up in production of these leaves and flowers, and consequently, at the end of summer, this part becomes shrivelled and dead, like a. The green leaves, however, after they have developed, manufacture a considerable amount of food, and this descends from the leaves, and is stored in the short stem, which thickens in consequence and becomes a new corm at the end of the season.

The buds in the axils of the leaves of the new corm remain near its apex, carry on the production of a new series of flowers, leaves and corms in the following year.

A corm, instead of possessing only one bud at its summit, as at h, often has several buds there, each of which develops into a new corm in the manner described; thus one corm may give rise to many in a single season.

A bulb often resembles a corm in external appearance, but consists of a comparatively small stem, upon which is arranged a number of thick,· fleshy, scale-leaves, which overlap each other

more or less completely. The whole structure is practically a huge bud, and in the axils of some of its scales are small, rudimentary buds. Familiar examples are met with in the onion, tulip, lily, hyacinth, snowdrop, and narcissus.

The onion seedling, develops several leaves during summer and the plant swells at its base and forms a bulb. A section reveals its structure. Tracing the leaves from the green parts downward, it is observed that the bases, especially of the inner ones, are thickened, and it is these leaf-bases which form the main mass of the bulb, the stem upon which they grow being comparativély small. At number of thick, fleshy, scale-leaves, which overlap each other more or less completely. The whole structure is practically a huge bud, and in the axils of some of its scales are small, rudimentary buds. Familiar examples are met with in the onion, tulip, lily, hyacinth, snowdrop, and narcissus.

The onion seedlingdevelops several leaves during summer and the plant swells at its base and forms a bulb. A section reveals its structure. Tracing the leaves from the green parts downward, it is observed that the bases, especially of the inner ones, are thickened, and it is these leaf-bases which form the main mass of the bulb, the stem (s) upon which they grow being comparatively small. At the end of summer, the green parts of the leaves die and shrivel; their lower parts, which have become thin, act as a cover for the rest of the bulb, and prevent the rapid loss of water from the interior.

The onion bulb, if planted next year, forms adventitious roots from the base of the stem, and the terminal growing-point inside grows up into the air, and produces leaves and an inflorescence of white flowers at the end of a long hollow stem. The buds in the axils of the scale leaves develop usually in the same manner, so that from one bulb several flowering shoots are often produced. The materials stored in the bulb-scales are used up in this development of the flowering stems, and after the production of ripe seeds, the whole plant is generally exhausted, and dies away, in which case the onion is a biennial plant.

Occasionally, however, some of the lateral buds from the axils of the scales do not produce inflorescences, but leafy shoots only, which form small bulbs in the same manner as an onion seedling. After the death of the parent, these smaller bulbs remain, and carry on the growth in the succeeding year. The onion plant in this instance is a perennial.

A tulip bulb in autumn consists of a short, thick stem, upon which are placed a series of large, overlapping, fleshy scales. The latter are complete leaves, and not merely leaf bases, as in the onion. At the apex of the stem is an embryonic shoot, having leaves upon it, and bearing a terminal flower; in the axils of some of the scales are rudimentary buds.

In spring the flower-bearing stem grows from within the bulb and comes above ground, carrying with it the flower and two or three leaves. This development takes place at the expense of the food stored in the scales: the latter therefore soon become soft, and at the end of the season shrivel up and decay.

The leaves on emerging from the soil turn green, and during the spring and summer manufacture a consfderable amount of food; that part of it not needed for the plant's immediate requirements is transferred to lateral axillary buds below ground, and is there stored. These buds consequently grow rapidly and become young daughter bulbs; one of them is shown at n in course of development.

Bulbs like the onion, tulip, and hyacinth, which have broad, concave scales arranged in such a manner that the outer ones completely enclose those within, are known as tunicated bulbs. In lilies the bulb scales are not so broad, and are arranged to overlap each other like the tiles on a roof: such bulbs are said to be imbricated.

Chapter 3

The Leaf

As previously noted leaves arise in all cases from buds, and are side or lateral appendages of the stems of plants. They may be of many forms but are generally flattened structures, and, with the exception of those known as floral leaves, usually have buds in their axils. Their growth differs from that of the stem and root in being of short duration, for after reaching a certain size their increase ceases.

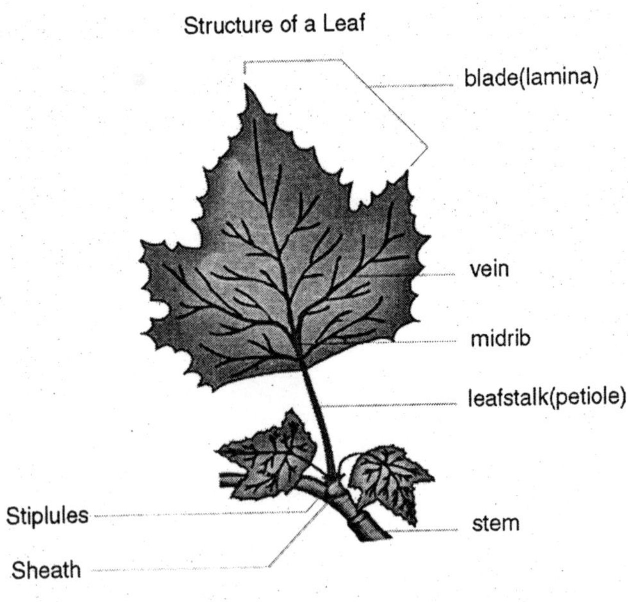

Fig. Leaf Structure.

FOLIAGE-LEAF

Those which are most conspicuous upon plants are green and are designated foliage-leaves. They are important organs generally concerned with the manufacture of food needed by the growing part of the plant, and are also organs from which much of the water taken from the soil by the roots is given off into the air.

A typical green foliage-leaf consists of the following parts—

1. A broad expanded portion termed the blade or lamina
2. A slender stalk or petiole;
3. A somewhat flattened basal sheath which connects the leaf to the stem.

The leaf-sheath often bears two appendages—the stipules—which may be broad and wing-like as in the clovers and pea, or small and narrow as in pear and apple; leaves possessing them are said to be stipulate, while those without are exstipufate.

The parts of the leaf are of very varied form. In the grasses the sheath completely embraces the stem, and in the Umbelliferae it is very prominent; in many plants it is scarcely visible.

The petiole where present is usually narrow and cylindrical; frequently it is very short or missing altogether, in which case the leaf is described as sessile.

The blade is generally the most obvious part of a foliage-leaf and the points of importance to notice at present are its venation, outline, margin, apex, and character of its surface.

Venation of Leaf-blade

The substance of the leaf is traversed by a number of woody strands which are termed veins or nerves, although it must not be inferred that they are similar in structure or function to the veins or nerves of animals. The arrangement of these strands is termed the venation of the leaf, of which there are two common types, namely

1. Parallel
2. Reticulate or net-venation.

In the first type the chief strands all run parallel to each other from the base of the leaf to the tip, as in the leaves of grasses, onion, hyacinth, lily-of-the-valley, and Monocotyledons generally.

In net-veined leaves the smaller very delicate strands form a fine net-work within the leaf and this arrangement is characteristic of Dicotyledons. Of reticulate veined leaves two divisions are made according to the arrangement of the main strands. In one, the leaves have a central strand or mid-rib running down the middle of the leaf and from it are given off slightly smaller branch strands; such leaves are pinnately veined or feather-veined, those of the apple, plum, and peach are good examples.

In the other division each leaf has several strong strands which start from the base of the blade and spread across to its margins somewhat like the fingers of an outstretched hand; such a leaf is described z&palmatdy veined. Ivy, sycamore, and currant leaves show this type of venation.

Forms of Blade

The outline of the blade of the leaf may assume almost any geometrical figure. When it is much elongated and narrow as in grasses it is termed a linear leaf.

It may also be lanceolate as in the narrow-leaved plantain; ovate egg-shaped); elliptical; reniform or kidney-shaped; cordate (heart-shaped); sagittate (arrow-shaped); spathulate (spoon-shaped) as in the daisy ; and hastate (halberd-shaped) as in sheep's sorrel.

MODIFIED LEAVES

Structures are often met with upon plants which although they do not possess all the parts of a foliage leaf as just described, are, nevertheless to be regarded as leaves on account of their origin and position upon the plant, and also by the fact that they frequently bear buds in their axils, and under some circumstances may become changed into ordinary green leaves. Several of these modified leaves receive special names as indicated below, according to their position upon the stem, or according to their texture, colour, and other peculiarities.

Cotyledons or Seed-leaves

These are the first leaves which a flowering plant possesses, and are nearly always simple and entire, and without stipules.

Some coniferous trees (pines and firs) have seedlings with several cotyledons, but dicotyledons usually possess only two, while in monocotyledonous plants only one is present.

In the bean, pea, and vetch they serve merely as storehouses for the food upon which the seedling depends for its early growth. In the cereals and grasses generally, the chief work of the cotyledon is to absorb the endosperm of the seed, and transfer it to the growing-points of the young root and shoot; while in the turnip, mangel and many other plants they come above ground and carry on the work of 'assimilation, thus behaving as ordinary foliage-leaves.

Scales

These are usually thin membranous leaf-structures, generally brown, white, or yellowish in colour, and may be either complete leaves, or merely the sheaths and stipules of leaves the blades of which have not developed.

On the stems above ground they are often present as coverings to the buds of trees and shrubs, acting as a protection for the interior of the bud against frost, heat, rUin, and the attacks of insects. Scales are always present upon the underground stems of perennial plants, and vary much in size. Upon the rhizomes of couch-grass and potato, they are small and membranous, while the leaves of a resting bulb are large scales, some of which are thick and fleshy, and stored with food.

Bracts and Bracteoles

The leaves which occur upon the stem at points where the flowers and inflorescences arise are termed bracts and bracteoles. They are very variable in size, texture and colour. In some plants they cannot be distinguished from the ordinary green foliage-leaves except by their position; more often they are rudimentary leaves somewhat resembling scales. The' chaffy bracts surrounding the flowers of grasses are termed glumes. In Arum, Iris, Narcissus, and snowdrop, a large bract, termed a spathe, encloses the whole inflorescence.

The cup of the acorn and the husk of the filbert and hazel-nut are persistent united bracts.

Bracts are sometimes brightly coloured.

Leaf-spines

In the sloe and other shrubs and trees certain branches are found which have been modified into short, stiff spines. That the latter are branches or shoots is seen from the fact that they frequently bear small leaves and buds.

In some plants however, such as barberry, the spines are evidently not branches, but modified leaves, for buds and stems frequently appear in their axils, and in the barberry all stages of transition between an ordinary green leaf and a branched spine are frequently observable on the same plant.

Leaf-tendrils

In the vetch and pea the terminal leaflets, instead of being green, are modified into thin, threadlike structures termed tendrils. They are sensitive to contact and wind round any small object which they touch. In some plants, such as the vine and passion-flower, the tendrils are not leaves but modified branches.

LEAF-ARRANGEMENT

Although to a casual observer the leaves appear to be without any regular arrangement upon a plant, careful inspection shows that they are distributed on the stems in a very definite order, which is usually constant for each species.

In some, such as the sycamore, dead-nettle, and cleavers, two or more leaves arise at the same node of the stem. Each collection of leaves is then called a whorl and the individuals comprising it are always separated from each other by regular angular intervals. Thus, if two leaves are present they are half the circumference of the stem apart or exactly opposite each other, and not both on the same side; if three arise at the same node, they are separated from each other by regular intervals of 120 degrees, or one-third of the circumference, and so on for any number of leaves.

On many stems the leaves are not in whorls but scattered singly along it, only one leaf arising at each node: such an arrangement is spoken of as alternate or spiral. A line drawn from the bottom to the top of a shoot in such a manner that it touches the base of each successive leaf is a spiral. The distances between the leaves measured along the stem are variable, some being an inch apart, others two or more; their angular intervals apart are, however, as definite and regular as in plants with the whorled arrangement.

The divergence or angular distance is usually expressed in fractions of the circumference. In elm, Spanish chestnut and grasses, it is that is, the spiral in passing from one leaf to the next winds half round the stem. In birch it is "J" while in pear and plum the angular distance is of the circumference.

If any particular leaf in a row is selected and the spiral traced round the stem touching each successive leaf until another leaf is reached on the same row, the number of leaves touched, not counting the one at which we start, is equal to the number of the denominator of the fractions expressing the angular divergence, and the numerator indicates the number of complete turns round the stem which the spiral line traces. For example, the angular divergence of the leaves on a pear shoot is £: selecting any one leaf as a starting point, the spiral line passes twice round the stem by the time that it reaches the next leaf on the same row, and in doing so touches the bases of five leaves. To determine the leaf-arrangement upon any particular shoot, it is necessary to observe the bases of the leaves and not the blades, as the position of the latter is affected by external conditions, especially by light and the force of gravitation. Occasionally the stems become twisted during growth, and the leaves are consequently displaced from their normal position.

The orderly arrangement of the leaves upon stems is de pendent on the internal forces of the living plant. By growing in this manner all the leaves become equally exposed to light and air, and interfere very much less with each others requirements in this respect, than would be the case if the leaves were disposed irregularly.

BUD-ARRANGEMENT

As buds arise normally in the axils of leaves, it follows that the arrangement of buds upon trees in winter will be similar to that of the leaves. A careful recognition of the position and arrangement of buds upon the shoots of plants is of some importance in the practice of pruning, where buds are required to produce branches growing in some particular direction.

LEAF FALL

In most of the broad-leaved trees and shrubs of temperate regions the leaves produced from buds in spring usually last only one growing-season, and then all fall off before the plants enter a state of rest iii the following winter.

A number of shrubs and trees, however, appear clothed with green leaves at all times of the year. These are described as ever-green. In such plants the leaves produced from buds in spring are not shed in the following autumn or winter, but live sometimes several seasons before they die and fall off. The length of time during which the leaf remains on a so-called evergreen tree after it is produced depends upon the kind of tree, the climate, situation, soil and other conditions.

In privet the leaves often remain on the twigs during winter, and fall off when the new buds open in spring; while in some conifers the leaves are not shed until they are ten years old or more.

The leaf usually separates from the shoot bearing it, at a point close up to the latter, and a more or less conspicuous mark, termed the leaf-scar, is left upon the shoot. The dangerous effects of an open wound is prevented by the formation of a protective layer of cork over the surface of the scar, which layer originates some time before the actual fall of the leaf.

Leaf-fall is not merely the dropping off of dead, withered leaves, but a distinct physiological process, which does not take place in leaves which are prematurely killed by the action of frost or excessive heat Moreover, the leaves do not fall off from branches of trees and shrubs broken or cut off, in early summer.

Although our attention has been chiefly directed to their morphology or origin, form and relationship to each other, it may be remarked that the work which these organs perform, for the benefit of the plant, is principally concerned with the maintenance of the life of the individual which bears them.

Sooner or later, however, flowers arise upon the plant, the special function of which is reproduction: in them seeds are produced containing embryos capable of developing into a new generation of plants when opportunity offers.

Before discussing the work of the flower it is necessary to become acquainted with the form and arrangement of its parts, and for this purpose it is advisable to begin with the study of a simple example such as a buttercup. In .the centre of the flower is seen a stem-like axis which is a continuation of flower-stalk. This is the receptacle of the flower and upon it is arranged a considerable number of lateral appendages of which there are four distinct forms present.

The lowermost of these appendages, that is, those farthest away from the apex of the receptacle, are yellowish-green in colour and resemble boat-shaped scales. There are five of them free from each other and arranged in a whorl: each is termed a sepal, and the whole collection or whort is known as the calyx of the flower.

Immediately above the sepals, and alternating with them, are five bright yellow heart-shaped leaves; these are the petals, the whole collection of which is termed the corolla of the flower.

Next to the whorl of petals are the stamens, of which there are a large number. Each consists of a thin thread-like stalk surmounted by a swollen and elongated tip. In the buttercup the stamens are not arranged in a whorl but in the form of a closely wound spiral round the receptacle.

Although the flower of a plant appears different in many respects from anything we have yet examined it is in reality a form of simple shoot or a stem with leaves upon it. The whole of its parts, however, have been modified to serve the purpose of seed production, and at first sight its likeness to a simple vegetative shoot is not appreciated.

That a flower is essentially equivalent to a simple shoot with very short internodes is, however, apparent from a study of its origin and position upon the plant and also from an examination of abnormal or monstrous flowers which occasionally occur.

A flower always occupies the position of a shoot; it arises either at the apex of a stem or in the axil of a leaf. Its receptacle, which normally ceases growth in length at an early period, occasionally grows on through the centre of the flower and develops into an ordinary leafy vegetative shoot.

The sepals, petals, stamens and carpels occupy the position of leaves upon the receptacle or axis of the flower; they are lateral appendages of the receptacle and are termed floral leaves. Moreover, the leaf-like character of the sepals and petals is generally obvious, and in so-called ' double flowers' some or all of the stamens and carpels assume the appearance of petals.

When the whole of the floral leaves are arranged in whorls, the flower is said to be cyclic: if they are inserted in a spiral line on the receptacle, the flower is described as acyclic. The term hemi-cyclic is applied to those flowers which like the buttercup have some of their floral leaves in whorls and others in spirals.

Generally the successive whorls alternate with each other: the petals for example are not opposite to the sepals, but occupy the spaces between the latter; the stamens alternate with the petals and the carpels with the stamens.

Very often the individual members of each separate whorl in a cyclic flower are all alike in shape and size; such a flower is regular while those in which this is not the case, as in the pea and violet, where some of the petals are larger than the rest, the flower is irregular.

All those flowers which can be divided into two equal and similar halves by a plane passing through the axis of the receptacle are symmetrical. Usually regular flowers can be divided into two halves by planes passing through the axis in several different directions: they are designated actinomorphic flowers, examples of which are chickweed, poppy and wallflower. Those which can be cut into two equal halves in one direction only are zygomorphic; for example vetch, pea and dead-nettle.

The number of members constituting each whorl in a flower is subject to much variation, but it will frequently be observed that in Monocotyledons each whorl consists of three floral leaves or some simple multiple of three (such as six or nine). In Dicotyledons the floral leaves are usually in fours or fives.

The pattern flower just described consists of four distinct kinds of floral leaves and is termed a complete flower. Sometimes flowers are met with in which one or more entire sets of floral leaves are missing—either calyx, corolla, androscium or gynaecium; such are spoken of as incomplete flowers: examples are seen in the mangel and ash.

THE RECEPTACLE

In the Buttercup the receptacle is an elongated cylindrical or conical axis and the whorls of floral leaves are arranged upon it at successively higher levels, the gynaecium occupying the highest and the calyx the lowest points respectively, with the corolla and androecium between. In many cases the receptacle is thicker and not so long as that of the buttercup, but the relative positions of the parts upon it is the same.

Flowers which like the buttercup have the corolla and androecium inserted on the receptacle at a lower level than the gynaecium and free from the latter are termed hypogynous flowers, and the gynaecium is described as superior; examples are charlock, poppy and chickweed.

In the plum the apex of the receptacle ceases to grow at an early stage, but the parts below the apex grow up all round it and form a hollow basin or urn-shaped structure, on the edge of which the calyx, corolla and stamens are arranged. The gynaecium, consisting of a single free carpel, is placed at the bottom of the hollow receptacle, this point being the real apex of the floral axis.

Flowers in which the corolla and androecium are arranged on the edge of a more or less hollow receptacle, surrounding the free gynsecium, are perigynous and the gynaecium is said to be superior as in hypogynous flowers. The flowers of plum, cherry, strawberry, are examples: in the strawberry, the part of the receptacle which bears the gynaecium is a solid lump, but round the latter the rest of the receptacle forms a flattish rim on which the petals and stamens are borne.

In some flowers the receptacle appears to be hollowed out as in the plum, but the carpels instead of being free from it are closely invested by its walls and completely adherent to the latter, so that the receptacle and gynaecium appear to be one structure: the ovaries of the carpels are imbedded in the receptacle, and only their stigmas and upper parts are free and exposed.

In such flowers the sepals, petals and stamens, seem as if they were produced on the upper part of the gynaecium, or its ovary, although in reality they spring from the receptacle which encloses and is completely united with the latter. Flowers of this type are described as epigynous and the gynaecium is inferior. Examples are seen in the apple, pear, gooseberry and carrot. The exact limits of the receptacle and the gynxcium cannot be seen or understood in fully developed flowers, and in many cases uncertainty exists in regard to them.

NON-ESSENTIAL PARTS OF THE FLOWER

THE PERIANTH

The calyx and corolla whorls of floral leaves together constitute the perianth of the flower, and as they are not directly concerned in the production of seeds are termed the non-essentialparts of the flower.

When one of the whorls of the perianth is absent as in the mangel, male hop, and anemone, the flower is spoken of as monochlamydeous; if both calyx and corolla are absent, as in the ash and willow, the flower is naked or achlamydeous.

The Calyx

The calyx forms a protective covering for the rest of the flower when the latter is still young, and may either lall off when the flower opens, in which case it is caducous^ or remain attached to the receptacle for an indefinite period, when it is described as a persistent calyx. It is usually green but may assume some other colour, in which case it is spoken of as petaloid.

A calyx which consists of free separate sepals, as in the buttercup, is termed polysepalous; those in which the sepals are united, as in the primrose and pea, are said to be gamosepalous.

In groundsel, thistle, and other plants belonging to the Compositae, the calyx takes the form of a ring of hair known as a pappus, which generally develops rapidly after the corolla has faded and acts as a float for the distribution of the seed-case by means of the wind.

The Corolla

This part of the flower is usually of bright colour and serves mainly as an attraction for insects. When the petals forming it are free from each other, as in the buttercup and rose, the corolla is polypetalous; the term gamopetalous is applied to corollas which are composed of united petals, as in the primrose and Canterbury bell.

THE ESSENTIAL PARTS OF THE FLOWER

The andrcecium and gynaecium are directly concerned in the production of seed, as explained hereafter, and are termed the essential parts of a flower.

The Andrcecium consists of stamens, each of which, is a modified form of leaf, although its appearance and structure is very different from the petals and sepals of the perianth.

A stamen usually consists of a more or less elongated threadlike portion — the filament — surmounted by a swollen thicker part termed the anther.

The anther consists of two somewhat elongated halves or anther-lobes, which are situated usually on opposite sides of the upper part of the filament: the part of the filament uniting the anther-lobes is termed the connective.

Most frequently the stamens of the androecium are distinct and completely free from each other as in the buttercup, but in some flowers the filaments of the stamens are united together and only the anthers are free. When all the filaments are united the stamens are described as monadelphous; if two or several separate bundles of united filaments are present the stamens are said to be diadelphous and polyadelphous respectively.

In the daisy, dandelion, and most plants belonging to the Composite, the anthers are united and the filaments are free; such stamens are termed syngenesious. Stamens attached to the petals, as in the potato flower, are described as epipetalous.

The gynaecium is composed of carpels, each of which generally consists of three parts:

1. A swollen hollow basal portion termed the ovary,
2. A thin more or less elongated part called the style,
3. The stigma.

Within the cavity of the ovary are small round or oval bodies termed ovules, which under certain circumstances develop into seeds. The part inside the ovary on which the ovules are borne is termed the placenta, The carpel may be considered as a leaf which has been folded along the midrib and united at its edges.

The line corresponding to the united edges of the leaf is termed the ventral suture of the carpel, and it is along this line that the ovules are generally attached in two rows—one row belonging to each edge; the line corresponding to the mWrib of the folded leaf is the dorsal suture.

These parts are readily seen in the pod of a pea, which bears considerable resemblance to a folded green leaf. The gynaecium may consist of separate carpels as in the buttercup, in which case it is said to be apocarpous. Frequently the carpels are united and then form what is termed a syncarpous gynaecium. The amount of union among the carpels varies, but very frequently their ovaries are completely united to form one common ovary: in such cases the styles are generally united to form one common style, the corresponding stigmas usually remaining free.

When the carpels of the syncarpous gynaecium are united by their edges as at the ovary possesses only one cavity or loculus and is said to be unilocular. In other examples the carpels are folded so that their edges meet in the middle of the ovary, the united parts forming partitions or dissepiments dividing up the common ovary into several cavities such ovaries are described as multilocular and each loculus corresponds to a single carpel.

Occasionally the number of loculi insftie an ovary does not correspond with the number of carpels present in the latter, as dissepiments occur which are not formed from the united walls of two neighbouring carpels but which are produced by the growth inwards of a portion of the ovary wall. The latter are termed false dissepiments, an example of which is the septum which divides the ovary in the Cruciferae.

PLACENTATION

The arrangement of the placentas or points from which the ovules arise inside an ovary is termed placentation. When the ovules are arranged in lines on the wall of the ovary, the placentation is parietal. In multilocular ovaries, such as at the ovules are generally arranged in the angles formed at the centre where the edges of the carpels are united, and the placentation is described as axile.

In the primrose and chickweed families of plants the ovules are attached to a placenta which arises in the form of a short column from the base of the ovary and has no connection with the sides: this arrangement is known as free central placentation.

MONOCLINOUS AND DICLINOUS FLOWERS

When both the essential parts are present in the same flower, as in the buttercup, charlock, and the majority of common plants, the flower is described as monoclinous ; sometimes the terms perfect, hermaphrodite or bisexual are applied to such flowers.

In certain flowers, as those of the cucumber, melon, hop, hazel, and willow, one or other of the essential parts are missing: such are said to be diclinous^ imperfect or unisexual. Diclinous flowers may be of two kinds:

1. Those in which the andrcecium is alone present and described as staminate or male flowers,
2. Those in which only the gynaecium is met with and spoken of as carpellary, pistillate or female flowers.

When both kinds of diclinous flowers are met with on the same individual plant, as in the case of the cucumber and hazel, the plant is said to be moncecious; in examples, such as the hop and willow where the two kinds of diclinous flowers are produced on separate individuals, the plants are spoken of as dioecious.

Chapter 4

Germination of Seeds

IT is well known that one of the most ordinary methods of raising plants is to sow what are called seeds, yet how few there are among the many who use them who fully appreciate their real nature and capabilities. This want of knowledge is not due perhaps so much to want of interest in them, as to the fact that for their proper management they are usually buried away in the ground, and are therefore unseen ; moreover, many of them are so small that their structure is difficult to observe with the naked eye.

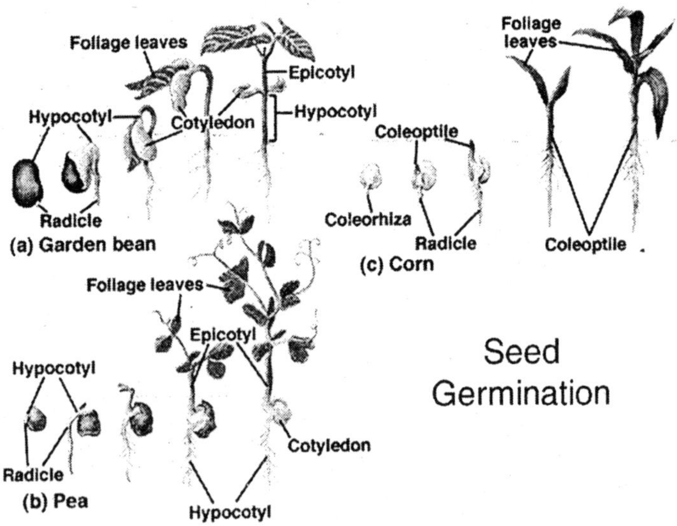

Fig. Seed Germination

In order to understand the true nature of a seed it is necessary to examine its origin and construction, and watch its development as far as possible from the earliest stages to the time when it gives rise to a completely formed young plant.

THE COMMON BEAN

A broad bean is one of the largest seeds met with in ordinary farm or garden practice, and as its parts are all sufficiently large to be observed without the special aid of anything more than an ordinary pocket lens, it is especially fitted for study.

When a nearly ripe pod of a broad bean plant is opened, each seed within it is found attached to.

The inside by means and it is through this stalk that all the nourishment passes from the parent to enable the young seed to develop. A1 first the pod exists in a rudimentary form in the centre of a flower and its parts and contents are very small; they are nevertheless readily seen with a pocket lens. After the fading of the flower, the pod and seeds within it grow larger and larger at the expense of food supplied by the rest of the plant, and ultimately when ripe the funicles wither and dry up, and the seeds become detached from the parent which has produced them.

When dry and ripe each bean seed is hard, with an uneven surface, but its internal construction cannot be clearly examined in this condition. On soaking in water for twelve hours, however, it becomes softer, and the parts can then be easily investigated.

The outside, which is a pale buff colour, is smooth, and has at one end a narrow elongated black scar called the hilum of the seed. It is known popularly as the ' eye' of the bean, and marks the place where the broad end of the funicle separated from the seed when it ripened in the pod.

Quite close to one end of the hilum is a very minute hole known as the micropyh, easily seen with a lens, and through which water oozes out usually accompanied by bubbles of air when soaked beans are squeezed between the finger and thumb. This opening communicates with the interior of the seed, and is the only one it possesses.

The rest of the seed after the testa is removed, is of oval flattened shape similar to the complete bean, and is divisible into two large fleshy halves called cotyledons which, however, are not completely separate from each other, but connected at the side with a conical projecting body, one end of which is found to fit into a hollow cavity in the seed-coat exactly opposite the micropyle; the other end is bent and turned inwards between the fleshy cotyledons. The extent and shape of this small curved structure is most easily observed when one of the cotyledons is removed completely; it remains attached to the other.

Soak some broad beans in water and keep them in a warm place all night. Examine them next day and make drawings of the various parts seen both before and after stripping off the testa. Observe the relative position of the parts of the embryo in reference to each other and to the seed-coat.

Examine and compare the structure of the following seeds after soaking in the same way:—Pea, scarlet runner beans, vetches, and red clover.

The bean seed contains nothing more than what has already been described; the nature and relationship of its component parts only become intelligible when the seed is placed in the ground or maintained under certain conditions, and allowed to grow. When growth commences the lower end of the small curved structure elongates and breaks its way through the coat of the seed at a point very close to the micropyle, but not, as often erroneously stated, through the micropyle itself.

It soon assumes the form and is recognised as a root of a young bean plant The upper bent half, which lies between the cotyledons, also pushes its way out of the same opening in the seed-coat and develops into a stem, from the tip of which leaves are gradually unfolded. It is thus seen that the seed of a broad bean is a packet containing a bean plant in a rudimentary condition

This plantlet is callai an embryo, and the portion of it which becomes root and stem is its primary axis. The part of the primary axis which is below the point where the cotyledons are attached consists of a very small piece of stem, the hypocotyl at the end of which is the radicle or primary root.

Where the stem ends and the root begins cannot be determined in the bean seedling without the aid of the microscope and examination of the internal structure of the axis of the plant.

The curved end of the primary axis above the cotyledons is the plumule of the embryo, and consists of a very short piece of stem, the tpicotyl on the top of which is a bud. From the latter is derived the ordinary stem which comes above ground with its green leaves and flowers.

In the early stages of the growth of the embryo from the seed the hypocotyl grows very little.

The upper part of the stem bearing the plumule comes out of the seed bent and it maintain this curved shape for some time after emerging. By this behaviour the delicate leaves of the plumule are protected from injury during their progress upwards when a seed is sown in earth or sand.

Fold up some soaked beans in two thicknesses of white flannel made damp, and place them on a plate. Cover them with another plate placed upside down, and leave them in a warm room. Examine them twice a day, leaving them, exposed to the fresh air for a few minutes each time, and keeping the flannel damp, not wet. When they sprout notice the place where the radicle has come out of the seed-coat. Let some grow till the radicle and plumule are well out of the seed, and compare the various parts of the sprouted seeds with unsprouted ones.

GERMINATION

When the pod of the bean is developing, the embryo in the seed is being fed by the parent and visibly grows until ripeness is attained. The young plant then assumes a dormant or resting state within the seed without showing any signs of life. Under certain conditions, however, the plantlet begins to wake up, and soon escapes from its protective coat to lead a separate and independent life.

This awakening from a resting condition to a state of active growth is called germination and is dependent upon an adequate supply of

1. Water
2. Heat
3. Air or oxygen.

It is also essential, of course, that the plantlet in the seed must be alive.

The exact nature of the dormant state of seeds is not understood, but in old seeds and those which are gathered in an immature condition or badly stored the embryo is often weakened or actually dead; in the latter case no germination is possible. The exact length of time which seeds may be kept before death of the embryo takes place has never been satisfactorily determined; it varies with the species of the seed, its ripeness and composition, and also with the method of storage.

In the case of most farm and garden seeds kept in the ordinary way, few of them are found capable of growth after ten years, and a large number die in two or three years. For present purposes it will suffice merely to mention that age is a determining factor in the germination of seeds.

Water is necessary is well known, as beans may be kept indefinitely in a sack or drawer at various temperatures and with access to air without germination taking place. When placed in moist ground, or between damp blotting-paper, they absorb water very readily. This is most easily observed when beans are soaked for twelve hours in a dish containing water. The water is transmitted through all parts of the coat, but much more quickly and easily through the micropyle and the line of softer material which runs the whole length of the centre of the hilum.

It is rapidly brought into contact with the part of the embryo which grows first, namely, the radicle. The soft spongy thicker part of the inside of the testa lying beneath the hilum stores up a considerable amount of water for the benefit of the developing plant, and the whole of the embryo and the seed-coat absorb water and become softer and larger in consequence; it is only after this swelling has happened that a bean begins to show any signs of germination.

The need of an adequate temperature for germination is a matter of common knowledge among those accustomed to sow seeds. If soaked beans are placed in the ground in midwinter they show little or no signs of waking from their dormant condition, yet when placed under a glass on damp blotting-paper indoors, the radicle makes its exit from the seed in a few days. Seeds differ in the temperature which is necessary to induce them to germinate, the embryos in som# commence to extend their radicles and push their way through the seed-coat even if just kept above freezing point; others require a temperature of 9° or 10° C. to start growth.

If attempts are made to grow beans at 45° C. it will be found to be too hot, and they make little or no progress. Between this high temperature at which growth appears impossible, and the freezing point where the development of the embryo of the bean is also suspended, there is a temperature at which the embryo makes the most rapid progress, and emerges from the seed-coat in the shortest possible time ; this most favourable temperature is about 28° C., both above it and below it the germination of the bean is retarded.

The supply of fresh air is also an essential condition for growth of the young plant from the bean seed, but the evidence for its need is not so manifest or so generally recognised as the necessity for moisture and warmth. It will be found, however, when beans are placed in a flask or bottle containing carbon dioxide or hydrogen gas they refuse to germinate, even when they are supplied with a proper amount of water.

The peculiar extension or growth of the parts of the interior of the bean seed, and the fact that a suitable supply of water, air and heat is necessary for the manifestation of these changes, suggests to us that we are dealing with a living structure. This becomes all the more«apparent when we observe that the oxygen of the air is absorbed, and in its place carbon dioxide is given off into the surrounding air, for this is what happens in the breathing of a living animal.

Carbon dioxide is produced when beans germinate. Place twenty soaked beans in a wide-mouthed bottle, and cork them up after showing that a match burns freely in the bottle. Leave them in a warm place for twenty-four hours, and try if a match will now burn in the bottle.

The carbon dioxide gas can be poured out into a beaker containing lime water; on shaking, its presence is proved by the lime water becoming 1 milky ' owing to the precipitation of carbonate of lime.

The particular use of the water, heat and air to the plant we cannot at present discuss. Without water the embryo would have little chance of becoming free from the tough and hard seed-coat surrounding it; water softens the latter, and makes it more easily torn by the extending radicle and plumule.

In the early stages of the life of the bean plant, from the commencement of germination up to the time when the first green leaves are unfolded, the development and building up of the elongating rootlet and shoot depend upon the thick cotyledons. At first the latter are thick and fleshy, but as the radicle and plumule grow the cotyledons become softer and thinner, ultimately shrivelling considerably. The cotyledons are leaves, the interior of which is packed with food for the rest of the growing embryo, and a large amount of the water absorbed by the seed is used for the purpose of dissolving the nutrient material in them, and carrying it from them to the various parts of the root and shoot of the young plant where growth is going on the cotyledons are essential to the development of the root and shoot of the embryo by cutting them off as soon as the two latter parts have emerged from the seed-coat.

Try separating one cotyledon and then two at various stages of development, and see if the axis (root and shoot) can be made to develop without them. The growth should be allowed to continue some time in order to obtain well-marked effects.

Not only do the changes observed innhe embryo of a germinating bean point to the conclusion that it is a living structure and like an animal dependent on a proper supply of water, heat and air for the manifestation of its life, but the parts of a young bean plant after emerging from the seed soon give evidence of the possession of peculiarities which are associated with life. When put in the ground, the radicle, in coming out of the seed, turns straight downwards and continues to grow in this direction.

This is the case no matter in what position the seed is placed. If, after germination has commenced, it is taken and replanted with the primary root pointing to the surface of the soil, the tip of the root soon begins to curve downwards again, and will maintain this course until again disturbed.

The plumule behaves in exactly the opposite manner; after emerging from the seed-coat its bent tip grows upwards and away from the root; if the seed is reversed and replanted the plumule begins to curve in such a manner that its tip is driven upwards towards the surface of the ground. That these peculiarities are somehow connected with life is clear, as dead embryos show no such behaviour.

Sow soaked beans in a flower pot or box filled with ordinary garden soil placing them in various positions in it, some laid on the flat side, some with the hilum directed upwards, and others with the hilum downwards. Allow them to grow in a warm place: take them up as soon as signs of germination are noticed, and observe the direction the root and shoot have taken.

The peculiar tendency for the root always to go downwards and stem upwards can be investigated by sowing beans in ordinary garden soil and afterwards reversing them. To avoid error all should be taken up, and then placed again in the soil in various positions—some as they were, a few with their roots and stems reversed, and others laid in a horizontal position. They may be re-examined at the end of a week.

When the roots have extended about half an inch take two seeds and suspend them by means of thread side by side in a bottle with their roots downwards and stem upwards. The bottle should contain a little water to keep the air damp. When the roots have grown about two inches reverse one of the seeds so that its root points upwards tad stem downwards. Notice that the tip of the root of the reversed seed in about twelve hours begins to turn downwards, while the plumule more slowly bends in such a way as to assume the position it had before it was reversed. The bottle should be placed in a dark box or cupboard to avoid the influence of light on the plant, and fresh air should be blown into the bottle twice a day.

Although seeds vary almost indefinitely in regard to size and shape they are similar to the bean in so far as they all contain a young plant packed away within the seed-coats. In this essential feature all seeds agree with few exceptions, and it is on account of the existence of a young plant within them that they are of use in the raising of crops or plants.

The manner, however, in which the embryo is arranged, and the relative size and appearance of its various parts, differ considerably in seeds; moreover, the growth during and after germination is not the same in all cases. A few of the more important and common variations in these respects must be noticed.

WHITE MUSTARD

The seed of white mustard (Brassica alba Boiss.) contains an embryo which like that of the bean consists of a radicle, plumule and two cotyledons; the latter, which are folded together, are relatively thinner than those of the bean and deeply notched. The radicle is bent round and lies in the fold of the cotyledons, between which is the very small, almost invisible plumule.

On germination the cotyledons, instead of remaining within the seed-coat and below the ground as in the case of the broad bean, escape from the enclosing coats altogether and grow up out of the ground, enlarging at the same time, and becoming green like ordinary leaves. They are the first ' smooth' leaves of the seedling mustard plant

After a short time the plumule grows up from between the cotyledons and forms a stem upon which are gradually unfolded the ordinary divided ' rough' leaves.

Soak white mustard seeds, and examine their structure, noting especially the way in which the embryos are packet in them. Allow some to germinate and grow for a week or more on damp flannel, and examine them in various stages of development, noting the notched cotyledons with small

The term hypogean is applied to cotyledons which remain below ground, those coming above being epigean, the relative amount of growth in the hypocotyl and epicotyl determining their position. If the hypocotyl grows vigorously during or after germination the cotyledons are forced above ground ; when only the epicotyl grows the plumule is lifted up above the soil, but the cotyledons remain below where the seed is placed.

In the broad bean the hypocotyl is very short, and the point where it ends and the root begins is not clearly denned. In a mustard seedling, however, the point of separation between the root and stem is somewhat swollen and readily distinguished.

All plants whose embryos, like those of the bean and mustard, possess two cotyledons, are known as Dicotyledons they form a very large, well-marked class of the flowering or seed-bearing plants.

The seeds contain within their coats nothing but an embryo plant, which depends for the development of its root and shoot upon the substances stored up in some part of its body, its cotyledons chiefly. This is true even in the case of seeds like those of white mustard, in which the cotyledons of the embryo are comparatively thin. There are, however, a number of plants, such as the ash, mangel and potato, which, although belonging to the Dicotyledons, have seeds in which there are stores of food inside the seed-coat, but free from the embryo and its cotyledons.

Such separate reserve-food is stored in that part of the seed known as the endosperm or 'albumen,1 and seeds in which it is present are called endospermous or albuminous seeds. Those like the bean, pea, and vetch, mustard, and turnip, which have no separate reserve-food, are known as exendospermous or exalbuminous seeds.

Take out a seed from the fruit of the ash tree in autumn ; carefully cut thin shavings from the flat side of the seed, stafting about the middle of the seed and cutting towards the narrow end. Note the white embryo with its well-marked radicle, hypocotyl and two flat cotyledons lying within the semi-transparent endosperm.

Some of the most commonly occurring endospermous seeds will be found to have embryos within them which are not dicotyledonous, and whose structure is in many respects. A good example is met with in the onion.

ONION

The seed is black, somewhat oval in outline, with one side convex, the other almost flat. Each contains within it endosperm and an embryo which lies curled up inside in the form seen. When germination commences, the curved part imbedded in the middle of the endosperm grows and forces the end of the embryo out of the seed. From this exposed end, which is the radicle, a straight, slender, primary root develops.

The part of the young seedling which extends from the root into the interior of the seed, grows very rapidly at first, at the same time assuming a sharply bent outline. It comes above ground in the form of a close loop but on further growth the end within the seed is pulled out of the soil, and grows up in the air. The tip within the seed changes and absorbs the endosperm, and usually remains there until all the nutrient material has been transferred from it to the various centres of growth in the young plant.

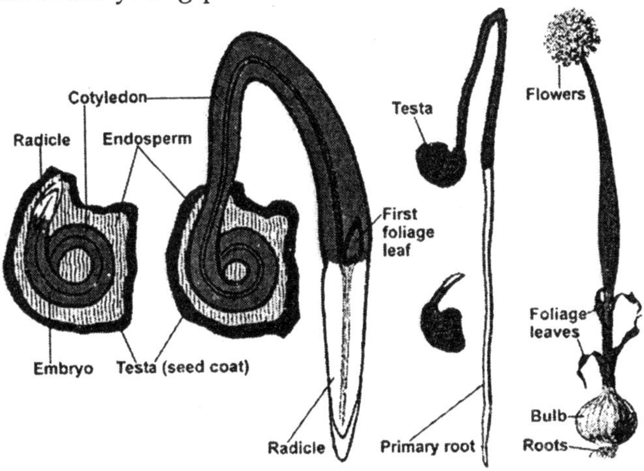

Fig. Onion Germination.

After the food-reserve is exhausted the tip withers and becomes free from the seed coat. In loose soils the latter is pulled above ground before the endosperm is exhausted, and remains on the end of the tip for some time. In other cases where the soil is damper and of a stiffer nature the seed-coat remains below ground altogether.

The curved part of the embryo which comes above ground is a leaf. It is the cotyledon of the embryo, and is in reality a thin hollow leaf like those of the full grown onion plant: within it is the plumule, which consists of a series of hollow conical leaves arranged one inside the other.

After one leaf emerges others soon follow, the younger ones coming out in regular order through slits in the sides of those immediately older than themselves.

Soak fresh onion seeds in water for a few hours. With a razor cut through some parallel to their flat sides in order to show the embryo within. Sow others in damp blotting-paper;

allow them to germinate and the seedlings to develop ; make observations of them at different stages of growth.

Watch the germination of seeds sown in boxes or pots containing ordinary garden soil.

Plants whose embryos possess only one cotyledon are known as Monocotyledons, and form the second large class of seed-bearing plants. Few of the representatives of this class with which we are ordinarily familiar have true seeds large enough for examination. The onion is probably one of the best commonly occurring examples which may be considered typical of the monocotyledons and easily obtainable.

To this class, however, belong all the grasses, but their seeds and embryos are so different in many respects from those of the onion that it is necessary to examine one of them in detail.

Wheat: A wheat grain, which may be taken as an example, is not a seed, but a kind of nut with a single seed within it. The seed grows in such a way as to completely fill up the interior of the nut, and become practically united with its inside wall.

The embryo occupies only a small part of the grain, the rest being taken up by the floury endosperm of the seed. The embryo is easily seen at the base of a soaked grain on the side opposite the furrow. When removed it has the appearance. The part of it which lies close up to the endosperm is a flattened somewhat fleshy shield-shaped structure called the scutcllum attached to the front of the scutellum is the plumule, consisting of a bud formed of an extremely short stem, upon which are sheath-like leaves enclosing each other.

The embryo generally possesses five roots, one primary and two pairs of secondary. They are all completely enclosed by a sheath which continuous wjtn the scutellum, and are consequently not visible from outside; their position, however, is marked by projecting bosses.

The sheath rc und the roots is termed the coleorhiza, and when germination takes place it expands and bursts the coats of the grain, the roots about the same time breaking through the enclosing coleo-rhiza. When a wheat grain is sown in the ground it remains there, but the plumule grows upwards, its first leaf, the coleoptile, appearing above the soil as a single pale tube-like structure; from a slit in the tip of the latter the first flat green blade soon appears, and is followed by a succession of single green leaves, the younger ones growing from within the older ones in regular order.

With a sharp knife or razor cut through from back to front, so as to divide the grain into two longitudinal halves, and note the floury endosperm and the shape and parts of the divided embryo.

Place a folded sheet of damp blotting paper on a plate, sow some soaked wheat grains on it, and cover with a tumbler. The grains will germinate. Watch their development up to the time the first green leaf appears, taking out the embryo and examining it at different stages of its growth.

There is difference of opinion as to which part of the embryo is to be considered the cotyledon. Soine authorities regard the scutellum as the cotyledon, while others give this name to the coleoptile or first sheathing leaf which comes above ground, and which has no green blade.

Others, again, consider that the first sheathing leaf is an extension of the scutellum, and the two combined is therefore the cotyledon. In any case, there is only one cotyledon present, and wheat therefore belongs to the class of monocotyledonous plants.

During the growth of the embryo of a wheat grain, it will be noticed that the endosperm becomes soft and decreases in quantity as the roots and plumule expand and develop; the endosperm is the food upon which the young plant depends during the early staggs of its life, the scutellum acting as a structure for changing, absorbing, and transferring this reserve-food to the growing parts which need it.

Remove the embryos from well-soaked grains, and grow them without the endosperm on damp blotting-paper. Allow ordinary uninjured grains to grow with them. Both the embryos in the grains and those removed from the grains develop, but there is a great difference in the results after a few days.

The store of reserve-food on which the young plant depends for its early development is sufficient to enable it to form a root, stem, and several leaves, as is evident when seeds are allowed to germinate upon damp flannel or blotting-paper, from which nothing but water is absorbed.

No food-materials or manures are needed for this primary development, and seeds germinate and the seedlings grow for a considerable time as well in poor soil or sand as in good rich ground. As soon as the reserve store is exhausted hunger becomes apparent, and unless the plants are then supplied with suitable nutriment from the soil and air, and are also placed under conditions favourable for growth, weakness and death are likely to occur.

Among the larger seeds, such as beans and peas, where there is an abundant store of reserve-food, the young seedlings begin to manufacture food for themselves from materials absorbed from the soil and air, long before their reserve is exhausted. In small seeds the reserve is sometimes almost consumed before the roots and leaves are sufficiently developed to carry on their work properly, in which cases a more or less temporary starvation and check to growth ensues. Especially does this happen when seeds are sown too deeply, for a large amount of food is then used in the production of a stem long enough to lift the leaves up into the air.

PRIMARY AND SECONDARY BOOTS

It was noticed, when dealing with the bean seedling, that its two ends always grow in opposite directions; the plantlet can be considered as an elongated axis, one end of which bears the leaves and invariably comes above ground, while the opposite end never bears leaves, and persistently follows the plumb-line downwards.

The descending part is known as the root. As will be pointed out later, all roots do not behave in this manner, and it is to be specially noted that many of the underground parts of plants are not roots; the exceptions, however, may be left for future consideration.

The first or primary root which the bean plant possesses is merely an extension of the radicle of the embryo which exists nrithin the seed. Soon after making its exit from the seed, it takes a downward course, and elongates by growth taking place near its tip.

Another point to be noted is that the lateral roots do not arise as up-growths on the surface of the primary root, but come from within it, and are described as endogenous. The slits which they make in the substance of the primary root, and through which they emerge, can be readily seen in a bean seedling. The secondary lateral roots are connected with its central more solid core; the three lowest ones, although they have just begun to grow, have not yet penetrated the outer layer of the root, and would not be visible on the outside of the latter.

This mode of origin is characteristic of lateral roots generally wherever they are met with.

Many dicotyledonous plants have roots similar to those of the bean plant. When, as in this case, the primary one continues to grow, keeping distinctly larger than the lateral ones, it is called a tap root. Very good examples are met with among cultivated plants in the carrot, mangel, red clover, and mustard in shep-herd's-purse, poppy, and many other weeds, as well as in most broad-leaved trees.

A number of plants have swollen fleshy roots in which food materials are stored for future use; they are described as tuberous roots, and must be distinguished from tubers, which are fleshy underground stems.

To designate the different forms of thickened roots various special terms are in use. The typical carrot root is conical; that of the turnip napiform. The root of the radish is spoken of as fusiform.

In some instances the primary root is soon rivalled in size by its branches ; it may even cease growth altogether. Such plants, on being pulled out of the ground, exhibit a bunch of slender roots, the chief of which are all much the same diameter and length ; roots of this character are described as fibrous and are well exemplified in common groundsel and grasses.

ADVENTITIOUS ROOTS

The roots of monocotyledonous plants differ in their development from those of dicotyledons. The single primary root of the onion, lasts but a short time, and is succeeded by others which do not arise as branches upon the primary one, but spring from the very short stem of the plant. Roots which arise on stems and leaves, or on various parts of the roots of plants, but not in acropetal succession are described as adventitious.

They are very common upon all monocotyledonous plants of the farm and garden, and may be considered the chief ones which such plants possess. In wheat, the embryo within the grain possesses three roots; in barley, five or six. Young barley plant show- use during the earlier stages of growth.

Although common upon monocotyledons adventitious roots are not confined to this class. Examples are abundant upon many kinds of dicotyledonous plants. Good instances are met with on the underground stems of mint, potato and hop, and on the runners of the strawberry, stems of creeping crowfoot and white clover, as well as many others.

They are generally produced at the joints where leaves grow upon the stem, and may arise in some plants from internal causes apart from any external influences; in other instances their development depends upon contact of the stem with water or moist soil. Almost all parts of certain plants may be made to produce them, and the propagation of plants Buch as gooseberries, currants, roses and hops by means of slips and cuttings depends upon their development. Pieces of stem cut off just below a leaf, and placed in moist earth soon develop adventitious roots near the cut end. Advantage is taken of their formation in the propagation of plants by means of layers.

Examine the roots upon any cuttings or slips which can be obtained, and observe whether they arise on the cut surface or at a point some distance away from it. Usually adventitious roots are thin and fibrous, but those of the dahlia are tuberous.

The complete root-systems of plants vary enormously in extent, but in all cases the total length is much greater than is usually anticipated. That of an ordinary oat plant, although not spreading through more than a cubic yard or two of soil, measured in one instance over one hundred and fifty yards in length.

A tree uprooted by the wind exposes to view a small number of stout roots similar to the thicker branches of the crown, and from these are given off a larger number of a finer texture. By far the greater bulk, however, which the tree possessed remain in the ground in the form of extremely fine rootlets or fibrils extending outwards generally as far, or a little farther, than the branches and leaves of the tree, but in some instances to much greater distances. Not only do the roots grow out horizontally and near the surface of the soil, but they extend downwards as well.

In isolated instances, where an adequate supply of air has been maintained along open cracks and fissures, roots have been found to descend many yards into the ground, but in general the roots of the tallest trees rarely go down more than seven or eight feet The want of air and presence of noxious substances in the lower regions of the soil checks further growth in this direction.

With many plants almost every cubic inch of soil immediately beneath their shade contains fine delicate rootlets, and the extent of their root-branching is very rarely realised on account of the ease with which these hair-like fibrils are broken off when the plant is pulled up or disturbed.

Many forest trees have a natural habit of sending their roots several feet into the soil Examples of fruit-trees belonging to this class, and requiring a deep soil for proper growth, are the cherry and wild pear.

Some trees, such as larch, keep their root-system nearer the surface, spreading out more horizontally in the ground. The quince, used as a stock on which to graft pears, has roots which remain in the upper regions of the soil. Similar surface-rooting habit is very marked in the ' Paradise' apple on which apples are grafted, and in Prunus Mahaleb upon which cherries are often grafted.

The root-system of wheat penetrates more deeply than that of barley; the mangel sends its fine rootlets more extensively into the deeper layers of the soil than cabbage or turnip red clover roots more deeply than white clover, and almost all plants have distinct and peculiar habits in this respect

The character and extent of root-development is not, however, altogether dependent upon the species of the plant concerned, but is materially influenced by external circumstances, such as the texture of the soil, and the amount of water in it. In deep open soils and loose sands the root-system of a plant is much larger than that of a similar plant grown in compact heavy ground.

In soils which are not water-logged, increase of moisture up to a certain extent increases the branching of the root, and excellent examples of the influence of water, coupled with good air supply, are seen in well managed plants in pots, and also among plants growing near wells and in drain pipes; the latter in some instances become completely blocked by the large number of fine rootlets of trees growing in their neighbourhood.

The root-system is also considerably modified by the totyl amount and kind of the manures or food-materials present in the soil. Up to a certain point an increase of nutrient substances increases root-development; an excess hinders it.

Mutilation influences the development of the root-system. If the growing-point of the tap root of a cabbage or tree is cut off its further elongation is prevented, but the secondary roots make up for the loss by more vigourous growth and often many adventitious roots arise near the cut end.

In order to properly cultivate plants of all kinds it is very important to study the habit or manner of branching of their roots and the relative proportions of the thick tap and secondary roots to the fine ramifications to which they give rise and which spread through the soil in all directions. The proportion of root-system below ground to the branches and leaves above is also worthy of attention.

The adaptability of plants to the various kinds of soils, their need of water, the cultivation which the ground should have and the rational application of manures to the plant are best understood and appreciated after a careful study of these points. Tap-rooted crops, such as sugar-beet, mangel, carrot, and parsnip, require the soil to be well-worked to some considerable depth.

Surface-rooting plants, such as barley, can be grown upon comparatively thin soils. The same is true of pears grafted on the quince and apples on the Paradise stock, and such plants respond more quickly to, and are more benefited by, top-dressings of soluble manures than plants with a deeply-penetrating root-system.

BOOT-HAIRS

On the root of a seedling bean grown on damp flannel or blotting-paper is seen a fine white silky band of hairs. These are called root-hairs, and they are never present at the extreme tip of the root but arise at some distance behind its growing point.

As the root elongates the root-hairs on the older parts die and turn brown, but others are produced on the younger parts, so that no matter what the length or size of the root may be for a short distance behind its tip it is clad with these delicate transparent hairs. When secondary roots make their appearance hairs are produced upon them in the same manner and follow the same order of development as those upon the primary root.

Their size and abundance are dependent on the species of the plant, and the amount of moisture surrounding the root. Plants growing in very wet situations or completely in water have few or no root-hairs. In very dry soils their development is checked, the greatest abundance being met with in moderately damp soils.

A good supply of lime is found to increase the number and length of the root-hairs of many plants. The hairs are hollow tube-like structures and should not be confused with fine small rootlets. They are outgrowths from the surface of the root, and as long as they last are concerned with the absorption of water and various ingredients in solution from the soil.

Upon plants growing in the ground the root-hairs are very intimately in contact with the particles of soil and are of so delicate a nature that it is practically impossible to remove a plant without destroying them.

Although almost invisible they are among the most important organs which plants possess. All the food constituents obtained from the soil and from the various manures applied to the latter are taken in by the root-hairs. By their means plants are kept constantly supplied with water; their destruction in the process of transplanting or any disturbance of their development and action, such as may be caused by excessive dryness or imperfect aeration of the soil, leads to a deficient water supply and consequent withering of the plant

Chapter 5

Growth Physiology

GERMINATION

The first inception of the living plant is from seed or spore. Subsequent developments based on the physiological processes of photosynthesis and respiration give rise to the mature individual. The birth of the new plant begins with the germination -of the seed or spore. The seed of the higher plants is a highly specialised structure consisting in the majority of cases of a resting embryo with first leaves or cotyledons, plumule and radicle and supplied with reserve food material upon which it draws in the preliminary stages of germination.

The spore of the lower plants, on the other hand, is no such specialised structure as the seed, and is incapable of withstanding prolonged periods of rest before the advent of suitable conditions for germination. The seed with its resting embryo can survive far greater environmental adversities than the less protected spore and is capable of germination on the arrival of suitable conditions.

The necessary factors for the germination of seeds can be classified under four main heads:—

1. Water.
2. Oxygen.
3. Temperature.
4. Light.

Water is an absolute necessity for successful germination. The quantity of water taken up by different seeds is variable ; wheat varies between 40 to 60 per cent., peas from 84 to 106 per cent., and in other plants such as maize, the value lies between 35 and 40 per cent, of the dry weight. The first stages of water absorption are due to imbibition. The cells of the embryo and endosperm are non-vacuolate, and it is only in the later stages of the process that cell vacuoles are formed and osmotic pressure comes into play.

It is still a matter of considerable controversy whether the soaking of seeds in water improves their germinating power. Morinaga has studied the capacity which seeds have of germinating under water in seventy-eight different genera, and has found that forty-three of these genera germinated under water. Seeds of two of the genera germinated better under water than on moist filter paper, while in eighteen of the cases germination under water was equal to that on damp filter paper. On the whole, it was discovered that small seeds germinated better under water than on filter paper. If the oxygen-content of the water wêre increased, several seeds, normally incapable of germination under water, germinated readily.

Temperature also plays a fundamental part in germination. If the temperature be too low, germination fails to occur, if too high, the seed may be permanently injured and killed. Temperature affects the rate of entry of water into the seed as well as the growth rate of radicle and plumule, it also affects the resistance of the seed coat to the extrusion of the radicle.

The temperatures at which different seeds germinate are different. *Phaseolus vulgaris* will show evidence of germination at 9° C., and the growth rate increases with temperature to 36° C. and ceases when 46° C. is reached. Rye will germinate at as low a temperature as 1° C., the maximum being 36° C. and the optimum 25° C. In all cases we are dealing with a minimum, maximum and optimum degree of temperature in germination ; although, as F. F. Black-man has shown, the " optimum " points are largely fictitious, varying with the time during which the factor has operated. The time-factor is of importance here, for the temperature may be too high for further growth to take place, onee the radicle has made its appearance ; in other words, for a brief period of time the high temperature has caused rapid growth, but the rate falls away owing to the operation of the time-factor.

Morinaga submitted the seeds of a number of genera to alternation of high and low temperature. As a general rule, each exposure to high temperature was for six hours, followed by exposure to a lower temperature for eighteen hours. Thus *Cynodon dactylon* germinated most readily with six-hour periods at 38° C., alternated with eighteen hour periods at 10° C., while alternation between 32° C. and 10° C. and 38° C. and 15° C. were less effective.

If oxygen be excluded, germination is inhibited, while certain seeds fail to germinate in the dark, and light is a requisite factor. According to Morinaga the seeds of *Typha latifolia* germinate more readily when the oxygen pressure of the air is reduced by partial replacement by hydrogen or nitrogen. The action of oxygen and light brings this discussion to the large and important question of the *dormancy of seeds.*

DORMANCY

It has already been stated that the seed of the higher plant is composed of an embryo which is differentiated into stem or plumule, radicle and cotyledons. The seed, as such, contains a minimum quantity of water, shows an absence of cell vacuoles ; and therefore in this condition chemical reactions, especially respiration, are reduced to a minimum. In such a state, the seed can retain its vitality over a prolonged period of time ; as much as fifty or eighty years in some exceptional cases. The seeds with the greatest vitality occur among the Leguminosae. As a general rule, seeds which show a high vitality possess a tough testa which prevents the absorption of water and is impervious to the action of gases.

The whole question of the factors affecting dormancy of seeds has been reviewed at length by Crocker, and an excellent summary of the physiology of dormancy is given in his paper. In general terms, Crocker found that dormancy or delayed germination can be attributed to one or more of the following causes:—

1. Incomplete development of the embryo.
2. Impermeability of the testa to water.
3. Mechanical restraint offered to the expansion of the embryo and other seed contents by the seed coats.
4. Inhibition and retardation of gases to, or, from the embryo, resulting in an accumulation of carbon dioxide within the tissues of the embryo, or in an insufficient supply of oxygen for germina-tion.
5. The necessity for the embryo itself, to undergo certain after ripening processes before germination and growth under ordinary conditions becomes possible.
6. Introduction by various means of a condition of dormancy in seeds previously capable of immediate germination.

To Crocker applied the special name "secondary dormancy."

According to Rose, freshly harvested seeds of *Tilia americana* with 10 per cent, or less of water-content, or seeds kept in warm storage for several months, fail to germinate when placed on a moist substratum at ordinary room temperatures. By numerous experiments he was able to show that the germination of the seeds could be brought about by a period of after-ripening in moist storage at 0° C. to 2° C., followed by a period of two or three weeks at 10° C. to 12° C. until germination was well started. To obtain vigourous growth of the seedlings a still higher temperature was needed. During the after-ripening process, it was discovered that the hydrogen-ion concentration of the seeds increased as well as the catalase and oxidase activity.

Rose also found that the freshly harvested seeds of *Sambucus canadensis* failed to germinate when sown at room temperature, but if they were kept in moist soil out of doors for the length of the winter, a high percentage germinated. He was unable to discover the factors influencing and controlling germination here. In *Rubus Idceus* he was able to show that dormancy was due to the high breaking strength of the endocarp. The presence or absence of light did not influence the germination of the seeds.

Pack ascertained that the seeds of *Juniperus* germinated readily if they were first maintained at a constant temperature of 5° C. for 100 days, to allow of after-ripening of the embryo. If the seeds, after the completion of the ripening process, were removed to a higher temperature (in this ease 15°) they were thrown into a state of secondary dormancy.

The following changes accompanied the after-ripening process:

1. An increase in the *pH* of the embryo ;
2. Increase of titratable acid ;
3. Decrease in stored fats and proteins ;
4. Increase in sugar-content;
5. Translocation of fats from endosperm to embryo ;
6. Slight increase in the rate of respiration of the embryo;
7. Increase in the value of the respiratory coefficient ;
8. Marked increase in catalase activity.

Kidd and West have shown that the seeds of *Brassica alba* sown in the presence of carbon dioxide could in certain circumstances be completely inhibited from germinating, and that this inhibition of germination might be maintained over prolonged periods ; as much as twelve months in their experiments, during the whole of which time the seeds lay on a moist bed of sand. When the seeds were exposed to air, Kidd and West discovered that this condition of secondary dormancy could in most cases be terminated by drying the seeds or removing the testas without drying.

No evidence of any changes in the seed coat could be found during this period of primary inhibition. On the other hand, it was shown that if the testas were removed with extreme care, the naked embryo remained dormant, and it was therefore suggested that a more stable condition of the tissues of the embryo had become established during this period of primary inhibition. Under the influence of carbon dioxide, embryos in this stable condition did not respond to normal environmental factors which govern the course of germination of ordinary embryos. Some very definite stimulus, either chemical or mechanical, was necessary to initiate cell division, and therefore the growth of the dormant embryo.

The condition of dormancy in the marine pea, *Lathyrus mari-timus,* has been investigated by Stiles and Bellow. The seed of this plant possesses a particularly tough testa which will not absorb water, even after immersion for a year, nor will the seed germinate when planted out. If, however, the testa be broken, germination proceeds normally. Stiles and Bellow suggested that in its normal surroundings, *L. maritimus* propagates itself either by vegetable suckers or that the seeds are cast on the shingle and the testas cracked by the action of the waves on the pebbles.

It is a well-known fact that willow *(Salix)* seeds cannot retain their vitality for more than a few days. Yozo Nakajima has shown that the seeds of *Salix pierotii* and *S. japonica* lose their vitality under normal conditions in seven days, but if they are maintained at 0° C. the power of germination is retained for a longer period. The seeds show still greater vitality if they are kept at 0° C. under rather dry conditions. In such circumstances germination with *S. pierotii* will occur 360 days after harvesting the seeds.

Davis found that the seeds of *Cornus florida* were dormant when the fruit has first matured, and that this dormancy could not be broken by treatment with acids, ether or ethylene. A period of after-ripening of 100 to 130 days was necessary at temperatures varying between 0° C. to 10° C. before germination would take place. During the after-ripening there was an increase in starch, sugar and amino-acid content of the seeds, but little change in fats, phosphatides or acidity. The catalase activity of the seeds was closely parallel to the after-ripening process. The seeds of *Sambucus canadensis* and *Berberis thunbergii* required alternating temperatures for germination.

Since in the seeds of *Cornus florida* increased capacity for germination was accompanied by increased catalase activity as well as an increase in the amount of starch, sugar and soluble protein, while the respiration fell until germination actually began, Davis suggested that the process of after-ripening was closely linked with respiration. Further evidence for this view was given by the fact that if the seeds were exposed to higher temperatures (15° C. and over) there was an increase in readily usable food material, the catalase activity fell and the respiration rose.

LIGHT

The action of light on germination has been much to the fore in recent years, and most of the work on this particular aspect of germination has been conducted in Germany. It has been known for a considerable time that the seeds of some plants do not germinate in the presence of light, while others refuse to germinate in the dark.

Busse found that the seeds of tobacco germinated more readily in the light than in darkness. Darkness, however, did not inhibit germination completely, but merely retarded it. Kinzel, working with the seeds of *Aquilegia atrata,* kept them on moist blotting paper for ten years. During the first year only 10 per cent, of the seeds germinated ; the remaining 90 per cent, lay dormant in a perfectly healthy condition for the whole period of ten years until exposed to light, germinated in fifty days.

Lehmann showed that the fully imbibed seeds of *Lyihrum salicaria,* of which only a fraction germinated in the dark, gave a germination of about 50 per cent, in twenty-four hours after a brief exposure (1/10 second) to a light intensity of 730 candle-power at a distance of 1 metre. Lehmann supposed that light has some kind of catalytic effect on seeds that will only germinate in the light, converting the insoluble reserve foods into soluble substances. In the case of " dark " seeds, *i.e.*, those seeds that will only germinate in the dark, the light was supposed to activate fluorescent organic compounds which have a destructive effect upon the proteolytic enzymes present.

Further work on *Lythrum salicaria* has been carried out by Lakshmana Rao, who allowed the seeds to germinate under constant conditions of temperature and varying light intensity. At higher temperatures (31° C. to 35° C.) with a single period of continuous lighting the product law was found to apply with very fair accuracy, that is to say, the percentage of germination was proportional to the intensity of the light and the time. At lower temperatures (25° C. to 28° C.) the action of the light was not so effective, and here germination was inversely proportional to the intensity of the light.

These variations from the product law are explained on the results of experiments with intermittent light. In this case the effect was proportional to the sum of the exposures used. With long periods of darkness, the effect of light exposure increased in proportion to the quantity of light used up to a maximum value, while with still longer dark periods, the period of light became less effective again, until finally the effect was less than with continuous light. Rao concluded from the results of his experiments that light had a direct effect on the embryo and not on the dead tissues of the testa.

Many organic and inorganic substances have the power of completely or partially counteracting the retarding influence of light on light-sensitive seeds. Kuhn, working with different seeds, studied the inter-relation of various mineral acids and light and darkness as factors influencing germination. Seeds of *Phacelia tanacetifolia* on filter paper moistened with distilled water gave a germination of over 80 per cent, in the dark, but only 18 per cent, when exposed to light. By the substitution of weak solutions of different acids, Kuhn was able to raise the germination to 60 to 80 per cent, of that in the light. In the dark, however, the acids appeared to be without effect upon germination. Similar results were obtained with *Solarium lycopersicum* and *Amaranthus atropurpureus.*

Magnus failed to find this encouraging action of dilute acids on the germination of *P. tanacetifolia.* A so-called " false germination " takes place by the extrusion of the embryo. He also found that an aqueous extract of the seeds tended to inhibit germination in the light, but, curiously enough, not in the dark. Further, if a portion of the seed near the micropyle were removed, good germination occurred in the light. It is thus evident that some substance is secreted near this portion of the seed that is responsible for the failure to germinate in the light.

The action of light on seeds is a complex one. Lehmann found that the seeds of *Veronica tournefortii,* which normally germinate in the dark, at a temperature of 30° C., did not germinate in light or darkness, while solutions of potassium nitrate stimulated germination in either light or darkness ; the results, taken as a whole, were better in the dark.

Treatment with potassium nitrate in the presence of light resulted in an 80 to 90 per cent, increase in germination as compared with the controls in distilled water. It was also found that the higher the intensity of the light, the greater the retarding effect upon germination, and potassium nitrate was quite incapable of counteracting this retarding influence. Thus with a light intensity of 440 candle-power at a distance of 1 metre and a temperature of 30° C., potassium nitrate had no effect upon germination.

According to Niethammer, with increasing age the sensitivity of the seed to light decreases. It would appear that the place in which the seed ripens plays no part in its sensitivity to light, and he concluded that under natural conditions sensitivity of seeds to light is an important factor in germination.

It has usually been supposed that the seeds of *Viscum album* (Mistletoe) require a prolonged resting period of five months before they are capable of germination. Heinricher was able by suitable experimental conditions of high relative humidity and light intensity of 1,600 candle-power to force the germination of the seeds at any time during the winter.

Boerher, in the course of an ecological survey of various forests, has come to the conclusion that light plays a distinct part in the problem of seed germination under natural conditions. In his opinion it has an indirect effect upon soil moisture, evaporation, and other factors.

The question of dormancy in relation to age, vitality and respiration of seeds and their oxidase and catalase content has been investigated by Crocker and Harrison. They were able to show that when the freshly harvested seeds of Johnson grass received a preliminary incubation at 20° C. in a germinator they remained dormant for a year or more. If a portion of the grain surrounding the embryo were removed, prompt germination occurred. It is evident then that dormancy here was brought about by the structures surrounding the embryo.

The catalase activity showed a marked reduction while the seeds were in the incubator at 20° C., a reduction of as much as 50 per cent, taking place in the first month. This fall in catalase content was accompanied by a fall in respiration; whereas there is normally a rise in such seeds kept under conditions of dry-storage.

If this gradual fall in respiration be of general occurrence in fully imbibed but dormant seeds, and, further, if the assumption be made that the death of such seeds is dependent upon the utilization and destruction of the reserve foods present by respiratory activity, this result must have an important bearing on the longevity of seeds under natural conditions. Brenchley found that many seeds would survive burial in the soil for as long as fifty-eight years.

MYCORRHIZA

Although in the large majority of cases temperature, oxygen, water-content and light or darkness are the principal factors concerned in germination, in a number of families of flowering plants germination is found to be dependent upon a fifth factor: the presence of a fungus. This state of affairs is especially prominent in the Orchidaceae and Ericaceae.

Unless the seeds of an orchid come into contact with its suitable fungus, no or very little germination occurs. It was on this account, although the scientific reason for the procedure was not understood, that orchid growers always sowed the seeds in soil in which orchids had already been growing. This association of fungus with higher plant, which is of wide occurrence in the vegetable world, was termed " mycorrhiza" by Frank. The mycorrhizal association of orchids has been widely studied, and now finds commercial application in horticulture. An excellent review of the whole of this subject has recently been published by Rayner, and only certain aspects will be considered here.

The classical work of the Frenchman, Bernard, has largely cleared away the difficulties surrounding the germination of orchid seeds. The seeds of orchids are minute dust-like particles, produced in enormous quantities from a single capsule, and each seed possesses an undifferentiated embryo.

The seeds prove to be quite sterile when sown in the ordinary way, but if the necessary fungus, here a species of *Rhizoctonia,* be present, germination proceeds rapidly and smoothly. In the case of *Bletilla hyacinthia,* Bernard found that in the absence of its fungus, germination was able to proceed to the first leaf stage, but this, as Ramsbottom pointed out, was probably a " last despairing effort " on the part of the seedling to survive in the absence of its fungus by the formation of its photosynthetic mechanism. In after-life the fungus is kept within the outer cells of the cortex of the root and never enters the endodermis or vascular cylinder.

An investigation of the *role* played by the fungus in the germination of orchid seeds has been carried out by Knudson, who has shown that the germination of the seeds of a hybrid orchid, a cross between a *Lcelia* and *Cattleya,* took place in the presence of sucrose solution and absence of fungus.

In the Ericaceae the problem of the relation between plant and fungus has been exhaustively studied by Rayner. Here, unlike the Orchidacea?, the fungal mycelium extends through the whole plant and infection of the seeds is brought about in the ovary. By sterilising seeds of *Calluna vulgaris* in weak mercuric chloride solution, Rayner showed that the root system failed to develop in the young seedling, and that such leaves as were formed soon showed marked chlorosis. Infected seeds, on the other hand, developed an extensive root system and grew normally.

This work of Rayner has been called in question by Christoph, who considered that the presence of the fungus was not necessary for the complete development of *Calluna,* and that growth proceeded as readily in sterilised as in unsterilised cultures. The fungus concerned here is a *Phoma,* a member of the fungi imperfccti. According to Rayner, *Vaccinium* presents a unique case among the Ericaceae and exhibits the most highly evolved symbiotic relation discovered in this family.

The fungus, as in *Calluna,* is distributed throughout the plant and extends to the ovaries and other organs of the flower. Rayner found it impossible to obtain seeds of *Vaccinium* free from fungus infection by sterilisation or removal of the testas previous to germination. All the tissues of the emerging seedling were subject to invasion by fungal mycelium, and it was probable that the normal germination of the seeds was bound up with this invasion.

The physiological relationship of fungus and higher plant is still a matter of controversy. It is probable that the mycorrhizal habit originated in a parasitic attack of the fungus upon the higher plant, and that the attack was at first repelled and later turned to advantage by controlling measures, with the final result that a fine balance of adjustment has been attained between the two organisms ; the fungus relieving the higher plant of certain of its metabolic functions.

Knudson held that the presence of the fungus facilitated the solution of insoluble carbohydrates until the assimilation mechanism had been firmly established. He found that *Rhizoctonia repens* induced germination of *Cattleya* seeds on a starch-containing medium and accelerated the germination of orchid seeds in a nutrient solution of sucrose. The fungus was able to effect germination *without infecting* the seed, due to its ability of digesting starch in the medium and modifying the *pVL* to a value favourable for growth.

On a peat and sphagnum substrate of nutrient solutions adjusted to a *pll* of 4-6 germination was as rapid as when the fungus was present; the especial factor for germination being that carbohydrate should be present.

In certain cases Rayner found that the fungus in *Calluna* could play the part of a parasite, especially if the seedling should happen to be at all weakly. The higher plant must therefore maintain a strict control over its fungal partner to prevent the latter from obtaining the upper hand and destroying the metabolic equilibrium.

The metabolic relations in after-life of fungus and higher plant is also obscure. The obvious statement to make in the circumstances is that the fungus must assimilate free nitrogen. Experiments, however, have only to be made on this subject to be immediately contradicted. More recently, Fulmer and Christensen have grown yeast on a synthetic medium of phosphates and molasses, and found that this plant was able to assimilate free nitrogen at 30° C. ; the assimilation was said to be a function of the *pH* of the medium.

No gain was registered for some six to eight weeks, and it was on this account that Fulmer and Christensen considered that many of the older workers failed to obtain positive results of nitrogen fixation with fungi. Nielson Jones and Llewellyn Smith have brought forward considerable evidence to show that *Phoma radicis,* the endophyte of *Calluna vulgaris,* does possess the power of fixing molecular nitrogen. By means of an apparatus entirely closed to the external air, errors due to the presence of ammonia and other gaseous nitrogen compounds were excluded, and the gaseous nitrogen used for the experiments was obtained electrolytically.

The amount of nitrogen fixed was of the order of 0-00095 gm. for 0-345 gm. of glucose used ; a value very considerably less than for the nitrogen fixing bacteria. In the absence of either combined or atmospheric nitrogen, the nitrogen value was 0-00016 gm. for 0-21 gm. of glucose used. Other workers have assumed that the fungus in some way assists in the absorption of salts and water from the soil.

Exactly how this may occur is difficult to see. McLeimen, from her studies on the endophytic mycorrhiza of *Lolium subulatum, L. perenne, L. multiflorum* and *L. temulentum,* found that there was a metabolic exchange between fungus and higher plant, with the result that the higher plant obtained a supply of fat or oil, rather than a supply of nitrogenous food material.

It is very probable that the metabolic functions of the fungus may be different in different families, and that in one case it may be concerned with the fixation of atmospheric nitrogen and in another with the production and exchange of fats.

GROWTH

Aided by assimilation and respiration, the seedling or sporeling is transformed into the mature individual. Growth is the final expression of successful metabolism. Without metabolism there can be no growth, and when metabolism ceases, death supervenes. It was early shown by Sachs that the only true criterion of growth is increase in dry-weight (material dried to constant weight at 100° C.), for transpiration and water absorption lead to considerable variations in fresh-weight, and no reliance can be placed on such figures.

Estimation of dry-weight, however, has the disadvantage that the plant is killed, and now increase in leaf area is much used to measure growth, for it has been shown that increase in this organ is also an index of true growth. Both the anabolic and the catabolic sides of metabolism are necessary for growth. The living plant must be supplied with food, and it must also be supplied with energy, without which its vital activities would soon cease. It is necessary, however, that the anabolic side of metabolism should be in excess of the catabolic, or no balance of material would be available for growth.

It has been shown by Boysen-Jenscn that a balance is always available under ordinary conditions for growth. He attempted to elucidate the economic working of two biologically different types of plants: *Sinapis,* a sun-loving plant, and *Oxalis,* a shade type. In *Sinapis* the maximum intensity of assimilation was estimated at 6 nig. of carbon dioxide per 50 cm.^{2} of leaf area every hour at 20° C. The dry-weight of a plant with a leaf area of 50 cm.^{2} was 0-279 gm., and the respiratory rate was found to be 0-8 mg. of carbon dioxide for the same leaf area and at the same temperature.

Boysen-Jensen was therefore able to calculate that the amount of dry-matter assimilated in the course of a day was 60 mg., and loss in dry-matter from respiration as 14 mg., or, in other words, the gain of material over loss was 46 mg., or 16-5 per cent. Here there is an ample margin of the anabolic over the catabolic side of metabolism to allow of fresh growth. In shade plants, such as *Oocalis,* the assimilation and respiration values are considerably lower than in *Sinapis.*

In *Oxalis* itself the maximal assimilation was 0-8 mg. of carbon dioxide per 50 cm.^{2} of leaf area every hour at 20° C., while the respiration rate varied between 0-1 to 0-2 mg. for a similar leaf area for the same time and temperature. Again, it is evident that there is a balance of material available for growth.

GROWTH FACTORS

External and internal factors have a powerful influence on the growth of the living plant. If the factors are not suitable, growth may be retarded or even completely inhibited ; or again, the absence of one or more factors may act in a detrimental manner on the well-being of the plant.

Among the principal factors concerned in growth are food and water. Without the right kind and quantity of food the plant cannot survive for any length of time. The whole subject of the food of plants is too large to be considered here in anything except brief outline. For a very full and detailed discussion of this subject, the treatise by Russell should be consulted.

THE ELEMENTS NEEDED TO BUILD UP PLANT TISSUES

Plant tissues are composed principally of the elements carbon, hydrogen, oxygen, nitrogen as well as sulphur, phosphorus, calcium magnesium and potassium. Besides these principal components, it is becoming increasingly apparent that particular elements are necessary in minute amounts to obtain normal growth in certain plants, and these appear to subserve the same purpose as the so-called vitamines in the animal economy.

The qualitative study of the mineral nutrients necessary for plant growth is usually carried out in water culture solutions by leaving out one of the necessary elements. The method has the inherent weakness that the physiological balance of the solution is disturbed and the question of antagonism enters into the problem.

Carbon

The only source of carbon for the higher green plants is the carbon dioxide of the air. This is present in the atmosphere in minute amount, approximately 3 parts in 10,000, but this low concentration is sufficient for the synthesis of carbohydrate. It has usually been considered that material increase of the carbon dioxide concentration has a toxic action on plants. This result, however, is due to the fact that the older workers employed highly impure sources of the gas, such as those from blast furnaces and gas flues. The carbon dioxide from such sources is always mixed with considerable amounts of sulphur dioxide and hydrogen sulphide; which are markedly toxic to plants.

Bolas and Henderson, using highly purified carbon dioxide and working under rigidly controlled conditions, found that artificial enrichment of the atmosphere with carbon dioxide resulted in a large increase in the dry-weight of cucumber as compared with the controls in normal air.

The increase became evident at an early stage of the growth, usually two to three days from the beginning of the experiment. In one experiment, with a concentration of 31-3 parts of carbon dioxide per 10,000, the plants showed a percentage increase in dry-weight of 60'6 18-5 over the controls with 3-9 parts of the gas in 10,000. These results will doubtless be of value in greenhouse horticulture in the future, and further observations should be of interest.

Certain of the lower green plants, especially the green algae of the soil, can function as either autotrophic or heterotrophic organisms in nutrition. Bristol Roach found that the soil alga, *Scenedesmus costulatus* var. *chlorelloides,* could be grown in a liquid medium of nutrient mineral salts alone in the light or with the addition of 1 per cent, of different carbohydrates in the dark.

With the addition of 1 per cent, glucose the organism was able to grow in the dark and retained its green colour. With maltose a similar result was obtained, but there appeared to be an initial " lag " period, preceding the straight line period of growth, while xylose was completely toxic to the organism.

In the light, glucose and maltose gave the best growth, while the order of other carbohydrates was: galactosc, sucrose and fructose. Other algae which were found to be heterotrophic were: *Cystococcus, Chlorella* and *Chlorococcum,* but they reacted very differently to the conditions imposed upon them. It is probable that it is not justifiable to regard the soil algae as a homogeneous physiological unit in consideration of their relation to soil fertility.

With *Scenedesmus costulatus* var. *chlorelloides* it was discovered that in high light intensity the organism was completely auto-trophic in nutrition, but as the light intensity was reduced, glucose was taken up from the medium. It is possible that when this alga occurs in the lower layers of the soil, it grows at the expense of certain organic substances that are directly available. When, however, growing on the surface layers of soil, provided that the moisture conditions are suitable, photosynthesis takes place at a rate dependent on the intensity of the light.

Nitrogen

The main source of nitrogen for the green plant is nitrates. Nitrogenous manures promote growth, their action is very swift, and they make the leaves a vivid green colour, causing rapid growth of leaf and stem. In large quantities they lead to rank and heavy growth and a tendency to susceptibility to infection from rust in wheat.

Hursch has found that the amount of sclerenchyma is reduced in proportion to the collenchyma in wheat with heavy nitrogen manuring, thus favouring the attack of *Puccinia graminis,* the mycelium of which can only develop in the collenchyma. Gregory has shown that barley increases its leaf area, but not its assimilation rate with addition of nitrate, whereas addition of phosphorus and potassium increase both assimilation rate and leaf area.

Appleton and Helms have ascertained that the rate of absorption of sodium nitrate by oats and cotton when applied at different stages of growth is more rapid the later the nitrate is applied. In both cases there is a close correlation between the rate of growth and the rate of nitrogen uptake. The effect of the amount of nitrogen supplied has been studied by Rippel and Ludwig on the growth of the sunflower ; for smaller rates of nitrogen-supply the actual rate of growth is less, but in the first half of the growth curve the relative production of dry matter and the relative absorption of nitrogen are greater with the smaller dressings.

Rippel and Ludwig have also attempted to correlate uptake of nitrogen with that of bases in the broad bean and oats grown in sand culture with or without the addition of combined nitrogen. The excess of nitrogen in the plants named above, which could have been absorbed in the form of nitrates combined with the bases present in the tissues (allowance being made for bases combined as sulphates and phosphates), is expressed as a percentage excess of nitrogen.

As would be expected, this figure is much higher for broad beans when combined nitrogen is withheld. In the case of oats, the excess is still greater when nitrogen is given, whereas nitrogen-starved oats show a large excess of bases. Turner has demonstrated the contrast between the response of barley and maize compared with flax to variations in nitrogen supply.

The former two crops show a large increase in the top to root ratio with increasing dressings of nitrate, which is due to a stimulation of the growth of the tops, and not to a depression of root formation. In the case of flax this effect is not very marked. Beyond a certain low level of nitrogenous dressing, further increases in nitrogen supply produce no appreciable change in the ratio of tops to roots.

Pearsall and Ewing have shown that plants grown in a medium containing large amounts of nitrates possess little protein, while low amounts of nitrogen lead to high protein-content. It is obvious, therefore, that the succulent condition of the tissues of plants grown under high conditions of nitrogen manuring cannot be due to imbibition of water by protein.

On the other hand, the ratio of protein to soluble nitrogen is always lower in plants receiving abundant nitrogen, *i.e.*, they have larger amounts of soluble nitrogen. The amino- and amido-nitrogen is always higher in high nitrogen manuring, and the chlorophyll content is also higher. The *pH* of the plant tissues under conditions of high nitrogen manuring is higher than with scanty nitrogen supply. This perhaps may be correlated with the fact that such plants as sorrel manufacture relatively larger quantities of organic acid, *e.g.*, oxalic acid, in soil deficient in nitrogen.

The water-content of the tissues with high nitrogen is always considerably greater than with low nitrogen. Pearsall and Ewing suggested that since abundant nitrogen leads to an accumulation of amino-acids, these influence the metabolism in such a way that the production of organic acid is reduced, leading to an increase in the *pll* of the tissues *(i.e.,* alkalinity). This high *pH* and high amino-acid content leads to a greater swelling of the protoplasmic colloids, and it is perhaps due to this feature that there is higher water-content and reduced transpiration of plants growing under conditions of high nitrogen manuring.

Phosphorus

The best source of phosphorus for plants is from phosphates. Phosphates promote root growth and have an important effect on the ripening of grain. The function of the phosphorus in the cell is difficult to ascertain. The presence of phosphorus appears necessary for mitotic division, possibly on account of the fact that phosphorus forms an integral part of the nucleus.

It is a well-known fact that the soil factors governing the supply of phosphate to the growing plant are considerably different from those concerned in the supply of most other nutrients. Von Wrangell distinguished three separate factors for the supply of phosphate from the soil:

1. The phosphate concentration of the soil solution.
2. The rate at which this concentration is restored after disturbance of the equilibrium between soil and soil solution.
3. The total reserve of soluble phosphate in the soil.

The first factor is considered to be of especial importance, because here one is dealing with easily soluble material as well as with substances of low solubility, the concentration of which largely depends on the presence or absence of other ions. MacGillivray has found that in phosphorus-starved tomatoes there is a re-utilisation of the phosphorus present; about half of the total amount is found in the fruit, irrespective of treatment, although, if there be a shortage of phosphorus, the size and number of fruits is much decreased.

Brenchley has investigated the influence of phosphate on barley at different periods of growth. The plants were grown in water culture with full phosphate allowance and also with no phosphate. The plants grown in full phosphate solution were deprived of phosphate after varying initial periods and *vice versa, i.e.,* plants grown in the absence of phosphate were supplied with the salt after initial periods of deprivation.

Provision of phosphate for the first six weeks allowed normal growth to take place, shown by the number of cars, tillers and grain produced, as well as dry-weight. Shorter initial periods of phosphate allowance led to serious shortage in these respects. If the phosphate were withheld for four weeks, tiller growth was not affected, but no ears were produced. With still longer initial periods of deprivation, growth was steadily depressed in all respects, and the usual bushy growth tended to pass over to the thin, lanky type.

The amount of phosphate absorbed by the plant increased more or less in direct proportion to the length of time phosphate was given at the beginning of growth, but sufficient was taken up in the first six weeks to allow of the plant making maximum dry-weight. The absence of phosphate in the early stages of growth, on the other hand, led to an extremely rapid drop in the ultimate amount of phosphate taken up by the plant.

It is evident that the importance of phosphate lies in the early stages of growth, and its application is of the most vital importance. Provided a sufficiency of phosphate be supplied in the early weeks of growth, tillering, ear, grain formation and increase in dry-weight proceed normally. These results are in contradiction to those of Pember and Pember and McLean, who found that barley can utilise a deficiency of phosphate at any time in the life of the plant.

This, as Brenchley points out, may well be due to the fact that they used well water in their experiments, which contains traces of phosphate, and, further, that they did not allow the tillers but only the main shoot to develop, so that such phosphate as was present was concentrated in its sphere of action.

Gericke, working with wheat, obtained the maximum dry-weight when the plants were grown in nutrient solutions for four weeks and then transferred to solutions containing no phosphate. Decrease occurred in the ultimate dry-weight as the initial period with phosphate was lengthened ; the plants receiving phosphate throughout the experimental period being amongst those with the lowest dry-weight. Brenchley never found this sharp fall in barley, although certainly there was decrease in the dry-weight with continuous supply of phosphate.

It would seem that the phosphate requirements of cereals at different times of development are by no means fixed or definite, but are influenced by various factors, such as environmental conditions (seasons, spacing of plants, time of sowing, etc.). Nevertheless, it seems to be a definitely established fact that the most critical period for phosphate nutrition is in the early stages of growth.

Potassium

Potassium aids in the production and translocation of carbohydrate, and appears to be of especial importance to leguminous plants. Maskell, in a series of observations on the production of starch in the potato, using a special technique which allowed observations to be made in the field, found that in plants growing on plots which had received respectively potassium chloride, potassium sulphate, low-grade potash salts and no potassium, statistical examination of the data showed that starch production was appreciably increased by potassium sulphate, but not by fertilisers containing potassium chloride.

The rate of translocation of starch from the leaflets on the potassium sulphate plots was also increased, but at the same time this varied significantly with other factors, of which, intensity of solar radiation and age were important.

Calcium

The absence of calcium causes stunting and discolouring of roots, as well as brown spotting and subsequent death of leaves. The older physiologists associated calcium with the formation of proteins as well as with the formation of the material composing the cell wall. Recent work has tended to confirm this view. Parker and Truog explained the association between protein and calcium as being due to the precipitation of acids formed as by-products in protein synthesis by this element; for example, the Leguminosac, which are rich in protein, are also rich in calcium. Newton did not support the conclusions of Parker and Truog.

He found that the intake of calcium by peas and wheat in water culture solution is not proportional to the nitrogen content. Newton explained the high calcium content of leguminous plants when grown in the soil to be due to the much larger amounts of carbon dioxide evolved by the roots of such plants in comparison with non-leguminous plants, and that this results in an increased uptake of calcium which does not necessarily bear any relation to the nitrogen content of the plants. The argument, however, is not entirely conclusive, and the subject requires further examination.

Neither barium nor strontium can replace calcium, as both these elements are extremely toxic to the well-being of the plant.

SECONDARY ELEMENTS IN PLANT NUTRITION

It has become increasingly apparent in recent years that besides the elements enumerated above as being essential in plant nutrition, other elements in minute amount are also necessary for the normal growth of the plant. The early pioneering investigations of Maze foreshadowed the possibility that, by more refined methods, the ten elements postulated by Knop and the older school of physiologists as satisfying all the requirements of plant growth might be insufficient. This view has been abundantly justified by a number of recent investigations.

The method of experimentation is to grow the plants in the usual nutrient solutions prepared from highly purified chemicals. In such circumstances many plants fail to grow normally. The addition of minute amounts of certain elements, e.g., zinc, manganese, aluminium and boron, has, in a number of cases, brought about normal growth.

Iron

Although iron does not form an integral part of the molecules of the chlorophyll pigments, nevertheless a small amount of iron is necessary for the production of these pigments in leaves. In 1920, Oddo and Polacci put forward the claim that iron could be replaced in culture solutions by the magnesium salt of pyrrole carboxylic acid. This statement, however, could not be substantiated by Deuber, who found that in no case did the substitution of the magnesium salt of this acid prevent chlorosis in cowpea, soya-bean, corn and *Spirodela.* In fact, in concentrations of 0-001 to 0-250 gm., the salt was definitely toxic to the plants.

Boron

It was shown by Warington that the broad bean, *Vicia faba,* only attained full development when grown in the presence of a trace of boron, and that no other element could replace it. In the absence of boron the plants showed a stunted development, and death eventually occurred in a very characteristic manner by the blackening and withering of the tips. The best results were obtained with amounts of the order of one part in a million, while larger quantities were toxic. Warington came to the conclusion that boron is essential for other leguminous crops, but not for barley. Brenchley and

Thornton have found that the nodules surrounding the roots of *Vicia faba,* which are responsible for nitrogen assimilation, are much reduced in the absence of boron. The vascular structure becomes defective, the strands being either entirely absent or only running for a short way into the nodule ; there is at the same time a reduced development of bacteroid forms, and the amount of nitrogen assimilation is much below normal, and, in fact, the bacteria have been found to change their symbiotic habit for one of parasitism, attacking the protoplasm of the host.

Warington has confirmed this work of Brenehley and Thornton, and found considerable hypertrophy of the cambial cells of *Viciafaba* in the absence of boron, as well as frequent disintegration of phloem and ground tissue and poor development of xylem. Brenehley and Warington have carried the matter a stage further, and have discovered that the need of boron is independent of the *pR* of the medium in which the plants are growing. Even an insoluble borate, such as aluminium borate, is effective in this connection.

Boron is essential for several leguminous plants as well as melon, whereas cereals arid candytuft can develop in its absence. The boron plays an integral part in the calcium metabolism of the plant; in its absence *Viciafaba* is unable to assimilate calcium.

Collings stated that, contrary to Warington's results, boron is not essential for the complete development of the soybean, although in water culture it does have a markedly stimulating effect. Sommer and Lipman found that boron is essential for the development of corn, peas, sunflower, vetch, barley, buckwheat, dahlias, lettuce, potatoes, millet, castor beans, sugar beet, kafia, sorghum, flax, mustard and pumpkin.

They found that dicotyledons respond more quickly to boron than do monocotyledons, and it may well be on ibis account that Warington failed to find any stimulating effect of boron on barley. The effect in certain cases may not become apparent for a week or two, or even for as long as a month (barley). They also found that zinc is necessary for wheat, buckwheat, broad beans and kidney beans.

Johnston and Dore found that boron is essential for the normal growth of the tomato. In the absence of this element the plants show four distinct types of injury:

1. Death of the terminal growing point of the stem
2. Breaking down of the conducting tissues in the stem
3. Characteristic brittlencss of stem and petiole
4. Roots extremely poor in growth and of a brownish unhealthy colour.

There appears to be evidence of a quantitative relationship between the amount of growth and the amount of boron present in the nutrient medium: the greatest amount of growth was obtained when the concentration of boron was 0-550 part per million.

Manganese

Minute amounts of manganese have also an important effect on the growth of certain plants. McHargue grew maize, onion, cucumber, lettuce, wheat, spinach, oats beans, tomato and peas in pot culture in quartz sand specially freed from manganese salts.

Two series of experiments were set up:

1. In which plants were grown with the addition of manganese carbonate
2. In the absence of any manganese compounds.

In the controls, when the manganese contents of the seeds, principally contained in the testas and embryo, were exhausted, growth ceased, and the leaves showed signs of chlorosis, and the starch and sugar content also fell. Normal growth was resumed when the manganese was added to the soil. Neither copper, iron, zinc nor arsenic was able to replace manganese in this respect, and McHargue considered it to be of equal importance to iron in the formation of the chlorophyll system.

Bertrand and Rosenblatt have found that leaves can be divided into four classes. In the first, which includes beet and *Aucuba,* the highest manganese content occurs in the youngest leaves. Later in the life of the plant the manganese content of the plant falls. In the second group, which includes box, ivy, iris and yew, the manganese content at first decreases and then rises to a higher value than before. In the third group, for example, *Syringa,* privet and chestnut, the manganese content increases at first very rapidly to a maximum, and this value then falls, while in the fourth and last group there is a steady increase in manganese content with age.

Further investigations of Bertrand and Rosenblatt have shown that in *Nicotiana rustica* arid *Lilium lancifoliwn,* the highest manganese content occurs in those organs which are most active biochemically. Thus, the flowers possess more manganese than the leaves and young leaves more than older leaves and green stems.

McLean claimed that manganese salts pumped into leaves *via* the stomata by means of a porometer are equally as effective as when they are added to the soil. Schreiner and Dawson have discovered that if tomatoes be grown in the field and also in pot culture, using a naturally calcareous soil practically free from manganese, the addition of parts per million of manganese sulphate produces normal growth in contrast to chlorosis in the untreated soils.

Silicon

The older physiologists considered silicon to be of importance in the nutrition of the Gramineae, and that its presence gave strength to the straw. Later experiments, however, have shown that silicates act by causing an increased assimilation of phosphate. Lemmermann and Wiessmann considered that perhaps silicates or colloidal silica could replace or partially replace phosphates.

This view has since been shown to be untenable, and Lemmermann himself has abandoned it. Lemmermann, Wiessmann and Sammet have now published the results of a further investigation, and have shown quite definitely that the favourable effect of silica is to be correlated with an increased assimilation of phosphates. Nemec has found that the amount of phosphoric acid taken up by the plant is proportional to the amount of water-soluble silicate in the soil.

Auximones

Bottomley considered from a number of experiments on the aquatic ferns, *Salvinia* and *Azolla,* and the common duckweed, *Lemna minor,* that, in addition to the ordinary culture medium (Detmer and Knop) made up of mineral salts, there was needed another factor for successful growth, and to this he gave the name *auximone.* By employing aquatics, Bottomley did away with the artificial conditions of growing normal land plants in culture solution.

He found it necessary to add organic matter to obtain healthy plants, and found that aqueous extracts of bacterised peat and material containing nucleic acids gave the best results. The amounts necessary were said to be minute, and these growth-promoting substances were considered to have a resemblance to the vitamines of the animal biochemist. Mockeridge claimed to have confirmed Bottomley's results, and Roseiiheim alleged that they held good for the growth of *Primula malacoides.*

This view of Bottomley on the necessity for subsidiary growth factors or auximones has not met with universal acceptance. Clark and Roller have shown that certain lower plants can be grown for months on end in a solution of purely inorganic salts, provided that the physiological balance of the solutions be correct. Similarly, Saeger has been unable to confirm Bottomley's results, and found normal growth of *Spirodela, Lemna* and other aquatics in ordinary culture solutions in the absence of organic matter, provided that the solutions were diluted ten times.

Wolfe has confirmed the work of Clark and Roller and Saeger, and suggested that the word auximone should now be dropped from the literature. Mockeridge, in a reply to these investigators, pointed out that their media may have become contaminated with bacteria which have a powerful effect on the growth of *Lemna,* and that it may be on this account they were able to grow these aquatics for extended periods in purely mineral culture solutions.

Ashby has now shown that the addition of organic matter does have a definite influence on the growth of *Lemna.* As Ashby pointed out, Bottomley and his co-workers made no attempt to control external conditions, especially the jpH of the medium. This latter factor is especially important, as *Lemna* is very sensitive to changes in *pH.* Ashby grew *Lemna* under rigidly controlled conditions in dilute culture solutions of mineral salts alone, and also with the addition of organic matter (aqueous extract of horse dung).

It was found that the *Lernna* could grow in culture solutions of pure mineral salts indefinitely, but the addition of organic matter produced certain remarkable results. The cultures were grown under constant conditions of light and temperature, and in constantly circulating air. Statistically examined, the results showed that the plants growing in cultures with the addition of dung extract (0-2 part per million) showed an increase in frond area and cell size over the controls.

The number of chloroplasts per frond also increased, resulting in an increased rate of photosynthesis and therefore in an increased growth-rate. The influence of the organic matter is apparently catalytic in nature, for increase in the amount of organic matter added in greater amount than 2-0 parts per million caused no increase in the growth-rate. It would therefore appear that though organic matter is not essential for the growth of *Lemna,* its addition does have the effect of increasing the growth-rate to a remarkable extent, thus confirming Bottomley's work.

TEMPERATURE

Since both the physiological processes of carbon assimilation and respiration are accelerated by a rise in temperature, and again, since assimilation is always in excess of respiration under normal conditions of'metabolism, it follows that an increase of temperature will lead to an increase of growth. The range of temperature suitable for growth is restricted. Too high a temperature leads to a collapse of the metabolic machine, and brings about the death of the plant.

According to Leitch, the effect of temperature on the growth-rate of *Pisum sativum* is shown by a uniform curve for the range of temperature—2° C. to 29° C.—which closely resembles the curves obtained for respiration by Kuijper. Above 29° C., marked fluctuations make their appearance, and no single curve can express the relationship between growth-rate and temperature.

Hence, for each rise in temperature a separate curve must be employed to express the rate of growth in successive periods of time. F. F. Blackmail's " time-factor " is of importance here. At 30° C. and 35° C. the rate of growth in the first ten minutes is the highest attained, while in the first half-hour there is a rapid fall, which in turn is succeeded by a second maximum showing recovery, and thereafter there is a gradual fall.

At 40° C. decrease in growth is rapid and uniform, and no recovery is registered. The coefficient for a rise in temperature of 10° shows a distinct falling off as the temperature increases, and from Leitch's results it is only between 10° C. and 20° C. that the

coefficient value lies between 2 and 3, so that the complete curve is not a Van't Hoff curve.

The greatest growth occurs at 30-3° C. Leiteh distinguished four main points from her results:

1. The minimum temperature,
2. The maximum temperature,
3. The optimum temperature,
4. The maximum rate temperature.

The minimum temperature is the lowest temperature at which growth takes place, the maximum temperature is the highest temperature at which the process will occur, while the optimum temperature is the highest temperature at which no time-factor is involved, and the maximum rate temperature is that temperature at which the process attains its highest intensity.

The variations discovered by Leitch at higher temperatures may well be due to a number of factors, the powers of activity of which have been increased at unequal rates. The solubility of different food materials, the stability of various enzymes that are present in the cells, and the accumulation of the end-products of respiration are all affected in different ways by an increase in temperature.

Priestley and Pearsall, in a critical examination of Leitch's results, pointed out that the growth of radicles is dependent upon the chemical reactions involved in the cells of the growing-point. According to these investigators, increase in temperature will increase both growth and hydrolysis, but the tissues of the growing point quickly use up the food immediately available, with the result that a decrease in growth must obtain for that period of time required for the food materials in their journey down to the apical regions of the root.

The arrival of this food is alleged to account for the second maximum in Leitch's curve. Finally, growth is diminished and brought to a standstill by the disorganisation of metabolism. This explanation certainly has the merit of ingenuity, but Leitch's results can be explained by any number of hypotheses should sufficient thought be given them.

Luyten, Joustra and Blaauw found that the earlier or later flowering of the Darwin tulip is dependent on the storage temperature. Bulbs stored at a temperature of 25° C. to 28° C. flower at the usual time ; if, however, the bulbs have been stored at lower temperatures, the period of flowering under field conditions is accelerated by almost three weeks. Again, flower formation in the tulip is possible at a very much lower temperature than in the hyacinth.

Gregory has found that the rate of increase in area of cucumber leaves grown under continuous conditions of artificial light falls off with the first measurement of area. Gregory considered that the " detrimental factor " concerned here may well have been the high temperatures which had to be maintained during the course of the experiments, which tended to increase the rate of respiration ; and he put forward three suggestions to account for the falling off of the growth rate: —

1. The growth ceases through the incipient starvation due to high respiration values, the concomitant of high temperature.
2. The detrimental effect of high temperatures and low light intensity is due to a change in the distribution of assimilated material in the plant.
3. To a direct action on leaf growth or to the factors acting simultaneously.

In a second scries of experiments, Gregory grew cucumber under conditions of constant light intensity and humidity at the following temperatures: 63° F., 70-8° F., 76-8° F., 84-4° F. and 90-3° F. Growth was measured as leaf area and dry-weight.

This indicates that with high temperatures, the falling off of relative leaf growth rate steadily increases.

The foliage leaves alone show growth curves which conform to a simple parabolic function: –

Where A is the leaf area at a time *t*, A_0 is the leaf area at time unity, and *r* is the measure of the growth rate. At sub-optimal temperatures the value *r* remains constant, but falls rapidly at supra-optimal temperatures. The relative leaf growth rate is thus independent of temperature after the unfolding of the first foliage leaf.

According to Gregory, at sub-optimal temperatures relative leaf growth rate is independent of temperature, but is controlled by light intensity. Reduction of the light intensity or shortening the period of illumination leads to a smaller leaf area. The temperature affects leaf growth at sub-optimal temperatures by its action on the developmental rate in the apical bud, thus controlling the rate of unfolding of the first foliage leaf.

At supra-optimal temperatures a time-factor operates which tends to reduce relative leaf growth rate. This is due in part to a redistribution of material in the plant, more growth relatively by the stem, and, in part, to the direct effect of high temperature on the cell divisions in the leaf primordia, leading to a reduction in the number of cells produced before and after expansion.

At all temperatures employed in these experiments and with the light intensity used, there was a fall in the relative leaf growth rate with time. A second time-factor must therefore be considered as operating due to the low light intensity. If light alone be operating detrimentally, this time-factor should disappear with increase in light intensity, and this was found to be the case.

LIGHT

In the vast majority of cases light is a large factor in the growth of plants. The effect of light on growth is fundamentally different from that of heat. A number of fungi, can pass through the whole of their life-cycle in darkness, while in the greater number of land plants the root system is intolerant of light. In general terms, it can be said that light possesses a formative influence, and is also indirectly important, since it provides the necessary energy for photosynthesis.

Light has a strong formative influence on plant organs, such as stem and leaves, which are markedly altered morphologically by prolonged exposure to darkness. Apart from alterations in the colour of the green pigments of the stem and leaves, plants exhibit other characteristics when exposed to darkness ; peculiarities which are summarised in the term *etiolation.*

Etiolation phenomena are best studied in plants with abundant supplies of reserve food, such as the dahlia or potato, since darkness deprives a plant of its assimilating powers. The general characteristics of etiolated plants are prolonged internodes and a suppression of the leaf lamina which becomes scale-like in shape. The xylem does not reach the same development as that of the normal plant grown in the light, and the etiolated shoot is soft, sappy and weak.

According to Priestley and Ewing, the stems of *Vicia faba, Solatium tuber osum.* and *Pisum, sativum* develop a true primary cndodcrmis with Casparian strip when grown in complete darkness, in place of a starch sheath. This eiidodermis is considered to restrict the supply of nutrient sap necessary for the growth to tissues within the cndodermal cylinder and the apex.

The development of the endodcrmis in the dark is said to account for the excessive elongation of the stem, as only the cells at the base of the meristematic tissues which cap the end of the endo-dermal cylinder receive adequate supplies of food. Growth of this meristematic tissue adds to the length of the stem, and growth in length is favoured at the expense of the normal lateral growth of leaf and cortex ; the angles of the stem of *Vicia faba* opposite the main vascular bundles fail to develop, and the stem remains rounded in cross-section.

Occasionally it was found in *P. sativum* and *V. faba* that on prolonged exposure to darkness, the Casparian strip may fail to develop though growth continues, and in such circumstances there is a greater development of lateral leaf extensions. In *Phaseolus multiflorus* no endodermis develops in the dark, and therefore the stem remains rounded in outline. Priestley claimed that in *V. faba* a constant feature of the etiolated shoot is that the cortical bundles seem to divide relatively much earlier than in the non-etiolated plant. In the etiolated shoot the starch sheath is replaced by a primary endodermis.

A striking stain which is alleged to show this endodermis is Nile Blue. From the reactions of the layer to osmic acid, the staining with Nile Blue seems to be due to the presence of fatty substances. The cells below the meristematic region are very fully packed with starch.

Priestley concluded that the main morphological and structural features associated with etiolation are determined by a redistribution of meristematic growth at the shoot apex, following on the greater difficulty experienced by the meristem in drawing nourishment from the vascular supply, because when grown in the dark, the walls between the vascular strands and meristem are rendered relatively impermeable by the retention within them of protein and fatty material that form the surface of the protoplast.

Unpublished observations by the present author have shown that *V. faba* does not develop a primary endodermis with Casparian strip when grown in complete darkness. In its early growth the apical meristem of the etiolated plant is certainly packed with starch which gradually disappears, until finally the starch granules in the starch sheath itself also disappear, but the walls never develop a Casparian dot.

In any case, it is placing a heavy burden on a single layer of cells to claim that great morphological changes in the plant body are brought about by their presence, and it is difficult to conceive why the migration of fats should lead to these alterations in shape.

The plant is a complex metabolic machine, and, as Gregory has pointed out, the action of light " is not inconsistent with a hypothesis of a master photochemical reaction independent of carbon assimilation leading to the formation of a substance directly involved in leaf expansion." There is a minimum light intensity below which growth cannot proceed, and unless this level of light intensity is maintained, a time-factor makes its appearance, leading to continuous reduction in relative leaf growth rate.

This reduction of growth rate may well be due to a fall in the growth rate of the roots. In barley, certainly, root growth and leaf growth are closely correlated. This close correlation may be due to the fact that the leaf supplies carbohydrate which is important for the growth of the roots, and the roots supply nitrate which is necessary for the growth of the leaves. The action of hormones in plant physiology is only now beginning to be realised, and no doubt further investigations will lead to considerable extensions of our knowledge of plant metabolism in the future.

Short exposures to light have a marked effect on the development of the leaf lamina in etiolated plants. Trumpf exposed plants of *Phaseolus multiflorus* for varying periods of time, such as thirty minutes, two hours, four hours and twelve hours, to artificial light, and found that even the shortest exposure to light produced profound morphological changes.

The inter-nodes became much shorter and the lamina broader. In a second series of experiments the plants were illuminated for one, five, ten and thirty minutes, and closely resembled plants exposed to a longer period in light, with the exception that plants having the longest daily exposure (ten to thirty minutes), although having well developed laminas, showed no development of chlorophyll.

Thus, in changes of form, the quantity of light is the factor concerned, and its action is not produced indirectly on the assimilating system. Trumpf also ascertained that in light of different wave-length, the blue rays and, to a certain extent, the red rays favoured elongation of the stem, while the blue light markedly induced lamina formation. Priestley has confirmed these observations, and also found that completely etiolated plants show no signs of lateral leaf development arid always retain their plumular hook.

Chapter 6

Plant Cell

The growth of plant cells is interesting on account of the fact that the protoplasm is covered with a solid wall. Any irreversible increase of the cell size involves the growth in area of this wall, which consists partly of crystallized substances. The development of cell walls, especially in the cells of higher plants, passes through two different stages. As long as there is growth in area, they remain thin and flexible. This fine membrane is called the *primary cell wall,* and it is said to grow in area by *intussusception*. As soon as the final shape and size of the cell are reached, the membrane is strengthened by adding new layers to the primary wall.

During this second stage the cell wall grows in thickness by *apposition*. The combined layers are called the *secondary cell wall*. The thickness of this is often so large that the slender primary wall is overlooked or neglected, especially in materials of technical importance such as wood pulp, textile fibres, etc.

Indirect methods of investigation, such as polarized light and X-rays, indicate that the primary wall consists of a loose submicroscopic network of crystallized strands of cellulose or chitin, arranged in a *dispersed texture*. In the secondary wall the framework is much more compact. This compactness is obtained by a parallel arrangement of the submicroscopic strands forming a highly anisotropic *parallel texture*.

Two different types of growth in the area of the primary wall can be distinguished:

1. *Tip growth* of filamentous cells such as fungal hyphae, hairs (e.g. root hairs), latex tubes, pollen tubes, etc. In this case, only the tip of the cell is involved in growth, and no intussusception occurs away from the end. Sometimes the tip growth of very long cells (e.g bast fibres of ramie, which reach a length of over 20 cm.) continue after the formation of the secondary wall has already started in the middle body of the cell.
2. *Extension growth* of the cells in elongating tissues. In this case the whole length of the primary wall is said to increase in area by intussusception. There is an enormous bibliography on this subject, because the extension growth is stimulated by auxin, which was considered at the time as the hormone of the extension growth.

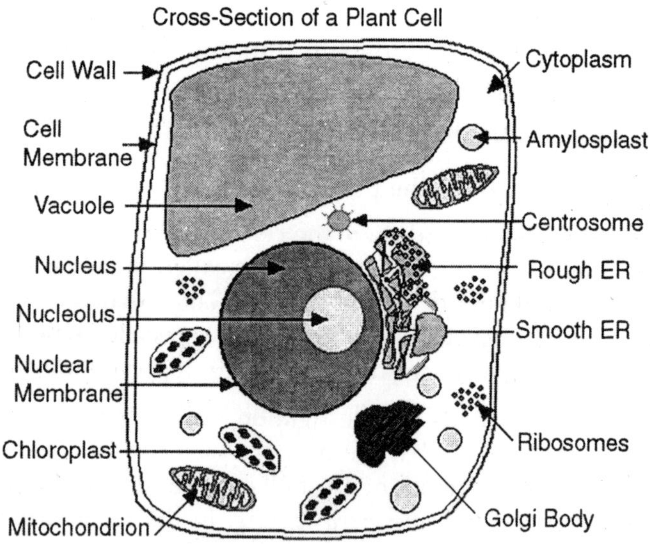

Fig. Plant Cell

However, the question how auxins intervene in this process cannot be solved until the mechanism of the growth in area has been cleared up. For a long time it was thought that the extension growth consisted of simple water intake, and the turgor stretch of the cell wall caused thereby. But it has been shown that this type of cell wall growth is much more complicated, since it involves respiration, biosynthesis, and morphogenesis. Moreover the work needed for the elastic deformation of the growing cell wall is negligible as compared with the total energy available for growth.

As extension growth concerns tissues with numerous cells which expand simultaneously at different rates, the geometrical problem how contiguous polyhedra can change their shape without being individualized arises. Krabbe thought that a disintegration of the tissue was inevitable and suggested the hypothesis of *sliding growth.* This implies that the cells of a tissue assume the individuality of independent unities growing along each other like unicellular protophytes. This is an extreme interpretation of sliding growth it is true, but it demonstrates clearly that such an idea is theoretically unsound.

Priestley published other arguments against the possibility of this type of growth; sliding of the two adjacent cell walls would interrupt the channels of pits, so that their orifices would no longer correspond. In consequence one might expect to find a large number of unilateral pits in tissues in which sliding growth had occurred.

As this is not the case, Priestley postulated that the adjacent cell faces must grow together at the same rate, and he termed this *symplastic growth.* His theory cannot, however, explain how latex tubes and other specialized cells grow across tissues and thereby come in touch with cells which previously were not their neighbours. This frequently occurs when derivatives of the cambial cells, such as tracheids and fibres, elongate in the tissue. Two adjacent cells are pushed apart by the tip of a cell which intrudes between them. This has been termed *intrusive growth* by Sinnot & Bloch or *interposition growth* by Schoch-Bodmer.

The last author emphasizes that only the tips of elongating fibres behave in this way. According to her, the walls of the central body of the cell do not grow any longer, so that the entity of the tissue is secured. Pits are only found in that part of the cell wall, and the fibre tips are free from them. From this description it follows that different parts of the cell wall must behave differently. A plant cell does not grow as a whole, but in a differentiating cell there are growing and full-grown parts at the same time.

STRUCTURE OF GROWING CELL WALLS

The electron microscope shows that young cell walls have a woven texture of dispersed cellulose microfibrils. The diameter of the microfibrils involved measures about 250 A. However, the electron microscope is said to be unsuited for research in cell physiology and growth because the specimens must be examined completely dried *in vacuo*. Nevertheless, it is possible to take advantage of the results obtained by electron microscopy if they are properly combined with those of other methods.

The natural state of the cellulosic framework of the living cell wall can be reconstructed from the electron micrograph if its chemical composition is known. In growing coleoptiles we have found 92.5% water. If this water content is taken as an average for all cell constituents, the cell wall contains only 7.5% dry matter. In corn coleoptiles 32 mm. in length twothirds of this dry material is composed of substances other than cellulose, such as pectins, hemicelluloses, wax.

Therefore, the growing cell wall consists of only 2.5% cellulose by weight. As the density of crystallized cellulose is about 1.55 this is even less by volume. It shows how astonishingly loose is this framework in a growing cell wall, including plenty of space for living cytoplasm and all kinds of incrusting substances. Only when this highly hydrated, spacious network is freed from non-cellulosic substances and completely dried does it furnish the picture of a dense interwoven texture in the electron microscope.

The impression of a rather compact framework is further enhanced by the high focal depth of the electron microscope, which yields a sharp picture of microfibrils situated in rather distant planes. Whereas it is difficult to understand how a woven texture as shown by the electron micrograph can grow by intussusception, where new microfibrils can readily be interlaced.

Another handicap to electron microscopy of cell walls has been the method of preparation, which previously consisted in disintegrating cells in a blendor whereby pieces thin enough for electron transmission were obtained. From these specimens it was not possible to locate the pieces examined in the original cell wall, and the micrographs obtained did not show whether the whole wall of a cell or only parts of it are involved in growth. Muhlethaler showed, however, that the purified cell walls of macerated meristems are thin enough for direct examination without cutting them into smaller bits.

This is possible because in these membranes he cellulose frame represents only about 2.5% by weight of the living wall, which has a diameter of less than 1¼ in the swollen state. So the dried cell wall freed of all incrusting substances would be only about 0.025¼ thick, if its mass were spread as a homogeneous film. Yet, according to the porous structure of the wall the dried framework is correspondingly thicker, say 0.075 ¼. This would correspond to a woven texture of about three microfibrils thick. As macerated collapsed cells consist of two superposed walls the thickness of the preparation may amount to 0.15¼, which is an appropriate value for the examination in the electron microscope.

The possibility of investigating whole cells proved an opportunity for observing the same spot of the wall alternatively in the electron and in the ordinary microscope. For this purpose the walls are stained with benzoazurin, and the resulting dichroism is examined in polarized light. As the direction of the strong absorption coincides with the main trend of the microfibrils, a mutual check on the observations in the electron microscope and the results of this indirect method is possible. Thereby an identification of different cell types and the investigation of individual faces of a cell wall are feasible.

TIP GROWTH

Our first intention was to elucidate the tip growth in order to observe the formation of new microfibrils at the very tip of the cell where intuesusception growth occurs. *Root hairs* seemed a favourable object for this purpose, because their tip growth is very fast. This research yielded the discovery that the tip of root hairs (*Zea mays*) is covered with a felt-like layer of cellulosic microfibrils. The felt is more than 1 ¼ thick and corresponds to the slime layer which covers the growth region of the hair. The microfibrils are anchored in the interwoven cell wall.

The slimy consistency of this layer is due to a large amount of interfibrillar substances with high swelling power, such as pectins. These can easily be demonstrated by pectin dyes (ruthenium red, methylene blue), whilst microchemical reactions, aiming at proving the presence of cellulose, fail. This is probably due to the fact that the microfibrils found are so well crystallized that no intermicellar spaces are left for the penetration of cellulose dyes.

The tip of the root hair is so woolly on account of this extracellular felt that the formation of new microfibrils cannot be observed. It is probable, however, that the cytoplasm which synthesizes the wall substances soaks this system of cellulose microfibrils, which on account of the hydrated state of the pectins must be very loose.

As the process of neoformation of microfibrils is hidden in the root hairs, we looked for objects where the growth zone is not covered by slimes, such as hyphae of fungi and pollen tubes. A favourable object is the growing *sporangiophore of Phycomyces* before the tip is inflated for the formation of the sporangium.

The chitinous cell walls of fungi have the same microfibrillar structure as those of higher plants; the only difference is that the microfibrils consist of chitin instead of cellulose. Curiously enough both types of microfibrils have approximately the same diameter and both are arranged in loosely woven dispersed or compact parallel textures. At the tip of the young sporangiophore the texture of the microfibrils is very dense, and is somewhat obscured by some incrusting substance which is difficult to remove.

The same difficulty is encountered with the tip of *pollen tubes*. These are very smooth and do not show any structure visible in the electron microscope. Only after extraction with such a strong leaching agent as 10% KOH can the microfibrils be made visible. Therefore, the new wall formed at the tip seems to be cutinized instantly in order to prevent desiccation of the pollen tube which is growing in air.

The microfibrils of pollen tubes have a smaller diameter than those of the microfibrils found hitherto in primary cell walls. The supposition that this might be due to the haploid state of the pollen tubes turned out to be wrong. There is no relationship between the size of microfibrils and the number of genoms in the nucleus. This was shown by looking at the haploid alga *Spirogyra*, where very coarse microfibrils occur in the cell wall.

Incidentally it was found that the rounded wall of the end cell in a *Spirogyra* thread showed an unexpected tip growth. At the very tip, longitudinal microfibrils spread out from the woven texture of the wall into the open, leaving a gap between them. Along the margin of this gap cross-fibrils are woven into the pattern of the cell wall. It seems that the cytoplasm of the cell tip is so to speak naked while forming the new wall.

EXTENSION GROWTH

The extension growth has been studied in the cells of macerated coleoptiles of oat and maize. This classical subject has yielded quite unexpected results. At a very young stage (oat coleoptiles of 5 mm. length) the *parenchyma cells* show an unexpectedly early formation of local secondary bands with parallel texture laid down on the woven primary wall. These reinforcements were found to be located along the edges of the polyhedral cells. It is difficult to understand how such a parallel-textured strip can extend when the cell elongates.

These strips are also visible in the polarizing microscope when stained with benzoazurin. Between the reinforced edges there are thin pitted areas. There again an extension seems quite impossible without distortion of the pits whose elliptic orifices have always a transverse orientation. The elongation of the cells is actually not due to a growth in area of the cell faces already formed, but to a polar or even *bipolar tip growth.* Since cells are rare in the macerated tissue, this tip growth must go on very rapidly.

A gap bordered by longitudinal microfibrils is clearly seen, as well as a margin where transverse microfibrils are woven into this warp. In the living state the tip seems to consist of cytoplasm only, so that the wallbuilding protoplasm protrudes beyond the formed solid wall. This observation excludes any active help of the turgor pressure in cell elongation, because this would force the cytoplasm out of the cell rather than extend the woven framework of the cell wall.

Cells with tip growth burst at the very tip when their turgor pressure is increased. This impressive phenomenon of plasmoptysis at the cell tip can easily be observed when growing pollen tubes are immersed in water.

Epidermal cells prove still more conclusively that cell extension is really tip growth. In oat coleoptiles these cells elongate from about 13 to 2000 ¼ (2 mm.), i.e. 150 times within 4 days. There is a very thick exterior wall which already exists before the cell extension starts and which holds the same thickness when the extension proceeds. The thick outer wall is visible as a broad parallel-textured band at the top of the figure, whilst the rest of the cell is tubular-textured as in the parenchyma cells.

At the cell tip this tubular texture grows, and the same manner of growth is used for the elongation of the thick parallel-textured exterior wall. There a much denser tuft of microfibrils is seen which is formed by the living cytoplasm from the cell. Thus the wall lengthens in its definite thickness by bipolar addition of parallelized microfibrils.

Sieve tubes and *tracheids* also extend by tip growth. For young fibres and tracheids this was already known. Whilst the cell has already differentiated to a ring tracheid at the bottom of the picture, the tip consists of a veil of interwoven cellulose microfibrils ending at the top in an open tuft of the longitudinal fibrils. It is clearly seen that the cytoplasm forms the warp first, and only afterwards is the weft plaited into the woven primary wall. Finally, the transverse microfibrils exceed those of the longitudinal warp in number.

By this type of growth, parenchyma cells, epidermal cells (apart from the outer wall), sieve tubes and the primary wall of tracheids obtain their *tubular texture* with transverse pits and the highest value of anisotropic properties, such as refraction, absorption, strength, etc., in the transverse direction as had been previously established by indirect methods such as double refraction and dichroism.

CELL-WALL GROWTH IN DIVIDING CELLS

How does the cell wall behave when the cell divides? In very young coleoptiles there is still some mitotic activity. We could therefore photograph a certain number of dividing cells. The structure of the expanding cell plate is difficult to observe, because it is inside the enveloping membrane of the mother cell. But the constriction of this wall in the middle of the cell is clearly visible. Its microfibrils are partly bent inwards, where they meet the cell plate.

Thereupon they are interwoven with the microfibrils of the corresponding lamella of the double-laminated cell plate. A certain number of the microfibrils of the original longitudinal wall are left over and decorate the margin of the two daughter cells with their submicroscopic fringes. These microfibrils are gradually dissolved and disappear. This proves that existing cellulose fibrils can be hydrolysed near by the place where other microfibrils are synthesized. Whether these two antagonistic processes are catalysed by the same enzymatic system is open to discussion.

The effort of our laboratory to elucidate the mechanism of extension growth has shown that, in the object investigated, there is no difference between tip growth and 'cell extension'. This had already been postulated previously. But it was thought then that the primary wall could expand by loosening its existing framework and subsequently inserting new microfibrils. In order to obtain a local loosening, certain microfibrils had to be dissolved by hydrolysis.

That a local 'melting down' of microfibrils is possible has been proved in this study, when we observed how the ring of microfibrils left over as fringes along the border of the new wall in divided cells gradually disappears. Thus, plasticizing of an existing cellulose frame by dissolution of microfibrils is possible. But in the observed cases of cell extension it does obviously not occur. Growth in area does not proceed by intussusception but by a terminal increment, just as the area of a cloth is enlarged by weaving its margin. In so doing the cytoplasm must somehow protrude beyond its solid case.

As the coleoptiles investigated represent the classical subject for the study of extension growth, there is some danger of generalizing these findings to all cases of growth in area of cell walls and of banishing the concept of intussusception. Although most elongating cells probably grow in the way here described, there are still some cases which must be thoroughly investigated before a general theory of cell-wall growth can be formulated. It will be very interesting to find out how tip growth proceeds in the very fast extension growth of grass filaments.

Further, there is an enormous growth in area of the wall when the descendant of a narrow cambial cell differentiates to form a large-bored vessel member. Here no tip growth is possible, but the cell surface must increase along some longitudinal line or lines, where plasticizing of the existing wall is necessary. The same is true when a cell produces some local outgrowth (e.g. a root hair). At the place where this reactivated tip growth starts, the existing framework must have been loosened somehow beforehand.

Since the texture of primary walls is so spacious, there are many possibilities of rendering the wall softer by increasing the interfibrillar ('incrusting') substances capable of swelling, or by intercalating new microfibrils.

This would be possible where living cytoplasm is located within the growing cell wall. Our study seems to indicate, however, that the cell wall is only soaked with living matter locally. Thus intussusception is restricted to small areas, and the cytoplasm retires very soon from the woven framework.

All these considerations show that the growth of the plant cell wall is an autonomic process. Exterior forces such as turgor pressure cannot play an essential role; they are only involved in so far as they influence the activity of the cytoplasm.

The extension growth of the cells in coleoptiles is a *bipolar tip growth*. The growing cell wall is living: it must be soaked with cytoplasm which synthesizes microfibrils *in situ*. In dividing cells it bends them to meet the cell plate on which they are fastened. Therefore the growth of the cell wall is an autonomic process which is governed by the morphogenetic faculties of the living protoplast.

THE MECHANISM OF PLANT CELL GROWTH

The plant cell after formation normally increases very considerably in size. This process may be studied in two situations:

1. When the cells are in intact organs,
2. When they are in isolated systems.

The first indicates the scope and characteristics of the process, the second the factors that control the several aspects of it. The two groups of studies based on the different situations complement each other, and it is therefore proposed to consider below, first, observations made with intact organs, and secondly, observations made with isolated fragments.

This treatment of plant celi growth is not intended to be exhaustive and one topic in particular is omitted. For two reasons no discussion is attempted of the nature of the auxin effect in cell growth. The subject is omitted, since it has recently been discussed in some detail by Audus and, secondly, since it is possible that an elucidation of this problem can only be developed after the nature of cell growth has been more fully explored.

Hitherto plant cell growth has frequently been described as cell extension. In our view this term is an unfortunate one. It is probably a legacy of the period when cell growth was considered a simple matter of an inflation due to the absorption of water. Recent work, however, has shown that the increase in size of the cell involves increases in many components of the system, and we therefore prefer the term cell growth to the older one of cell extension.

CELL GROWTH IN INTACT ORGANS

The process of cell growth involves the changes that occur between the meristematic and the mature fully vacuolated state. The meristematic cell is small isodiametric and without a prominent central vacuole. As the cell enlarges a central vacuole develops and the shape changes. Eventually a large parenchymatous cell is established with a thick wall and a thin peripheral layer of cytoplasm.

This set of changes is readily studied on the root. It is probable that this is a more satisfactory experimental object than the coleoptile, since with this organ the earlier stages of growth are not readily accessible to investigation. The root, being an organ of unlimited growth, has an apical meristem from which cells are being continuously generated.

These cells immediately they are formed commence enlarging, and thus at increasing distances from the apex the tissue is composed of cells in progressively advanced stages of development.

A similar situation is involved in the apex of the stem. Here, however, the position is complicated by the development of lateral members from the surface of the meristematic dome. In the root the surfaces of the growing regions are not covered with developing primordia, and this circumstance has been exploited in the development of two sets of techniques for the study of cell growth.

Since the surface of the growing region is naked in the root direct observations may be made microscopically on epidermal cells. The dimensions of epidermal cells may be measured at increasing distances from the apex and the growth of these cells determined over the whole range of development. This technique has been used in principle by Burstrom, Brumfield and by Goodwin & Stepka. It has been shown that sliding growth is not involved, and changes in surface cells may therefore be taken as representative of corresponding changes occurring in deeper tissues.

This general technique has yielded important results, and it has the merit that it involves observations on a more or less uniform tissue. It carries the limitation, however, that only a limited variety of observations can be made with it. It cannot, for instance, yield data regarding metabolic conditions in developing cells. In order to avoid this difficulty another experimental approach has been proposed by Brown & Broadbent.

These workers cut successive sections from the apex backwards, made observations on each section, and related the values obtained to the total number of cells in each section. They used a technique for determining the number of cells in a tissue originally developed by Brown & Rickless.

The range of observations that can be made with this technique is clearly greater than that possible with the earlier procedure. It involves, however, two possible sources of error. Cell division does not cease at the same distance from the apex in all regions of the root, and all cells in the section are therefore not exactly at the same stage of development. Secondly, sectioning necessarily involves a destruction of cells the extent of which cannot easily be estimated.

The first source of error is not decisive, since the distance over which cell division ceases is small relative to the total distance over which growth occurs, and the average given by the section therefore represents approximately the position for all the cells in the section. The error from the destruction of cells can be reduced by increasing the length of the section and thus decreasing the proportion of damaged cells.

These data are from the observations of Brown & Broadbent. The full development occurs over a distance of about 5.0 mm. from the apex and involves about a twenty- to thirtyfold increase in volume. The enlargement is accompanied by important metabolic changes. In the present connexion it may be noted that the enlargement in the early stages is accompanied by the gradual suppression of the capacity to divide. The zone that stretches from the tip to about 1.5 mm. behind it is that over which mitotic figures are normally distributed.

At the same time the observations of Wagner and of Gray & Scholes have shown that the mitotic frequency is greatest at the tip and decreases progressively with increasing distance from it. Brown has emphasized that the frequency of mitotic figures may not be an index of the relative rate of division. In this case, however, since the frequency decreases to zero it probably is, and the data therefore suggest that after formation at the apex the increase in the average cell size is accompanied by a corresponding increase in the average length of the interphase.

As growth proceeds the probability of division decreases, and at a certain point the probability becomes so small as to constitute virtual cessation of the process. The conditions that determine this transition have not been explored, but it is clear that the fact that it occurs indicates that growth in the early stages is accompanied by metabolic changes that tend to suppress division.

At about the point at which division ceases a prominent central vacuole appears in the cell. This vacuole, which is probably a development from a dispersed vacuolar system in the meristematic cell, increases in size with the enlargement of the cell. During growth, however, the percentage water content is increasing, indicating that the volume of the vacuole is increasing more rapidly than the other components, and that the enlargement of the cell is being determined primarily by the absorption of water. At the same time it is evident that the increase in water content is being sustained by relatively smaller increases in other phases of the system.

These data indicate that the dry weight of the cell increases about tenfold during the course of growth. Preston & Clark and Wirth have shown that the cellulose content of cells in the coleoptile increases during growth and Wirth has demonstrated corresponding changes with soluble carbohydrates. The increase in dry weight of the cell in the root is therefore probably due partly to similar changes. It is evident, however, that it is not due only to changes in these constituents.

In 1941 Blank & Frey-Wyssling described a series of important observations on cell growth in the coleoptile of maize. They reported an increase in protein content during the enlargement of the cell. Blank & Frey-Wyssling found similar changes occurred during cell enlargement in the hypanthium of *Oenothera*.

Kopp calculated that in the root the protein content increased, and Brown & Broadbent demonstrated this condition experimentally in pea roots. The appropriate curve is reproduced from the data of Brown & Broadbent, and it is evident from this that protein increases progressively during growth of the cell, resulting in at least a fivefold increase. At the same time it may be noted that when growth ceases there is a slight but significant decrease in protein, the final level being lower than it is when growth stops, although greater than it is in the meristematic region.

The increase in protein necessarily implies an increase in the general metabolic activity of the cell during growth. Kopp calculated and Brown & Broadbent later demonstrated an increase in respiration rate per cell during growth. The data of Brown & Broadbent suggest, however, that when growth ceases there is a slight fall in respiration which is coincident with the decrease in protein.

The increase in protein content might also be expected to increase the quantities of particular enzymes. We have recently been examining the changes in the activities of certain enzymes during the growth of the cell. These data have been obtained by taking successive sections along bean roots, killing the sections by freezing at - 10° C., determining on the killed tissue the activities of certain enzymes, and finally reducing the values for each section to a unit cell basis.

The activities of three enzyme systems have been investigated: a dipeptidase which hydrolyses alanylglycine, an invertase, and an acid phosphatase. The activity of the dipeptidase at increasing distances from the apex of the root of barley has already been examined by Bottelier, Holter & Linderstrom-Lang, and Wanner & Leupold have determined the relative sucrose-splitting capacity in different regions of the root of maize. In the bean roots growing in the conditions of our experiments growth continues over about the first 8 mm.

The data show that over this range there is about a fivefold increase in dipeptidase, about a twofold increase in phosphatase, and about a twentyfold increase in invertase activity. Further, it is evident that in about the region where growth in length is ceasing the activities of the respective enzyme systems decrease. Determinations of protein in cells of bean roots are not available, but in view of the earlier observations with peas it is probable that the increase in enzyme activity is due to the differentiation of part of the protein increment into enzyme bodies, and that the decrease in enzyme is similarly partly due to a decrease in protein.

It is an important aspect of the situation that the activities of the three enzymes do not change relatively to the same extent. The relative increases are different and the ratios of the activities of any two of the enzymes therefore change during the course of growth. A similar position is indicated by some observations of Linderstrom-Lang & Holter on the activities of what may be two distinct enzymes cleaving leucyl-glycine and alanylglycine respectively at increasing distances from the tip of the root of bàrley. The data, although they are not expressed in terms of unit cell number, nevertheless show that the ratio between the two enzymes changes as the distance increases.

Clearly since the ratios between the enzyme activities change it is evident that the metabolic pattern must change during the course of growth. During the growth of the cell a differentiation probably occurs in the metabolic state such that the relative intensities of different reactions change in the course of the process. In particular, these observations on the enzyme complement suggest that the metabolic activities of the young cell are different from those of the mature.

This conclusion is consistent with the results of an investigation by Morgan of the amino-acid composition of the proteins in different parts of the growing zone of the bean root. The proteins from sections taken at increasing distances from the apex of the root have been hydrolysed and the hydrolysates examined chromatographically.

It is evident that the composition of the proteins varies considerably. Such differences in the composition of the proteins suggest of course corresponding differences in the generalized metabolic pattern. This general conclusion is also consistent with a variety of observations by other workers. Berry & Brock have reported that the respiration of the tip of the onion root is more sensitive to cyanide than is that of the mature tissue, and Kopp has found that per unit protein it is lower at the tip than it is in the mature regions.

The data of Brown & Broadbent suggest that the increase in protein per unit increase in volume is greater in the earlier than it is in the later stages of growth. Finally, Burstrom has shown that the reactions of root cells to sugar, temperature and heteroauxin are different in the early and late stages.

CELL GROWTH IN ISOLATED FRAGMENTS

The elucidation of the mechanism of cell growth requires, of course, an analysis of the effects of different external conditions on the overall process and of each of its constituent aspects. Attempts have been made to develop such an analysis with intact organs. The interpretation of the results obtained in this experimental situation, however, is a matter of considerable difficulty, since the effects observed may not be due to a direct influence on the growing cells themselves.

In the intact root these metabolites may be supplied to the growing cell from mature tissue. Clearly in this situation differences observed on growing cells may not be due to primary effects on these cells, but may represent secondary effects from influences exerted on mature or meristematic cells.

This difficulty is avoided by studying the effects of different experimental treatments on isolated fragments composed of growing cells. This technique was first used by Bonner with *Avena* coleoptile tissue. This worker excised a 3 mm. fragment from the body of a coleoptile which had not reached the mature state, and subsequently observed the increase in length of the fragment in different media. A similar technique has since been used by Christiansen & Thimann with fragments from pea internodes and by Brown & Sutcliffe with fragments excised from the extending zone of the root.

This technique is an important one and has given valuable results. Nevertheless the limitations of it as it has hitherto been used require some consideration. It is an important feature of the technique that the fragment is normally excised from an organ that has completed most of its development. Thus, although the mature coleoptile may be about 4 cm. long, the fragment is frequently taken from an organ which is about 2 cm. long, and the cells of which would therefore normally only grow to about double their size. In the overall growth of the coleoptile the increase in cell size must be considerably greater than this.

Avery & Burkholder have shown with the *Avena* coleoptile from 4.0 to about 10 mm. growth is accompanied by cell division, and that from 1.O to 4.0 cm. growth is entirely by cell enlargement.

Thus after cell division ceases there is at least a fourfold increase in length, and in view of observations on the root it is possible that cell size is increasing in this phase also, giving an overall increase in size greater than fourfold. A similar situation is involved with the pea stem and root fragments. The root segments of Brown & Sutcliffe are taken from the zone which stretches from 1.5 to 3.0 mm. from the tip. The whole growth zone only occupies about 5.0 mm., and therefore at the time of excision at least the more mature cells in the fragment have completed most of their growth.

The maturity of the tissue from which the fragments are taken no doubt accounts at least partly for the limited growth normally given under experimental conditions. In favourable circumstances fragments from the coleoptile may double in length and those from the root quadruple. These increments, however, are considerably smaller than the normal twentyfold increase in volume with a tenfold increase in length that occurs when the cells are growing in an intact organ.

The limited increases observed are of course highly significant with respect to the final stages. On the other hand, in view of the results of observations on the changes in the enzyme complement during growth it is doubtful if results obtained with the final stages are applicable to the earlier.

Since the metabolic pattern changes as the cell enlarges it is of course probable that the conditions that limit growth are different in the different phases of the process. This has been demonstrated by measuring the growth during a 24 hr. period of successive millimetre fragments excised from the root of the bean. The fragments have been cultured with a 2 % sucrose medium on sintered glass disks.

It increases with increasing distance from the tip until about 3.0 mm. and then decreases. Clearly while it is the sugar supply that limits growth in the 3-4 mm. fragment, it is not this factor which controls the situation in more immature tissue. The growth after 4.0 mm. is less intense, since the cells are approaching the mature state.

Thus the evidence suggests that observations with the excised fragments as they have been used hitherto are relevant only to an analysis of the final stages of growth, and it is within this limitation.

Recent work has emphasized the importance of the metabolic state of the tissue in the determination of the course of cell growth. Frequently, however, the rate and total amount of growth are considered as being linked to certain metabolic processes in the sense that an acceleration of the metabolic process leads to a corresponding acceleration in growth, and, conversely, that an inhibition of the metabolic process depresses growth. Evidence is assembled here, however, which suggests that growth may be interpreted in terms of two complementary sets of processes. One set when accelerated tends to promote growth, and a second when accelerated tends to depress it.

Whatever the metabolic processes involved are, however, it is clear that they can only affect the growth process through a comparatively limited set of physiological conditions, and stimulations or inhibitions of growth must be interpreted immediately in terms of these conditions.

The increase in volume of the fragment is determined of course primarily by the absorption of water which as in any other vacuolated cell is the resultant of the conditions tending to draw water into the cell and of the restraining pressure exerted by the wall. Water may flow into the cell either along an osmotic gradient or as a result of a possible secretion into the vacuole which involves the consumption of metabolic energy.

The importance of the osmotic gradient has been emphasized by Thimann & Schneider, who have shown that when coleoptile fragments are immersed in a solution of mannitol which is not absorbed growth is depressed. In view of this it is evident that if growth is to proceed at a high rate the situation requires the continued absorption and accumulation of solutes in the vacuole, and the rate of solute absorption particularly when fragments are immersed in fluids of high osmotic pressure is clearly an important aspect of the situation.

The restraining pressure of the wall in the earlier phases of the investigation of cell growth was defined in terms of elasticity and plasticity.

These are no doubt important aspects of the situation. It is evident, however, that the expansion of the wall and therefore the restraining pressure is determined primarily by the synthesis and deposition of wall constituents. Bonner showed with coleoptile fragments immersed in water or in fructose solutions that during growth there is a synthesis of wall material, and Brown & Sutcliffe found with root fragments that extensive growth only occurs when cellulose is being synthesized.

At the same time it is probable that the synthesis of cellulose is only effective in promoting expansion when the physical state is such that the deposition of the cellulose in the wall can occur. Gorter has recently shown that in root hairs which have been treated with triiodobenzoic acid conditions are induced which do not affect the synthesis of cellulose, but prevent the intussusception of it in the primary wall.

Thus the growth of the fragment immediately depends on the operation of at least three and possibly four physiological conditions:

1. A possible secretion mechanism for water,
2. The absorption and accumulation of solutes,
3. The physical state of the wall
4. The synthesis of wall material.

These conditions are controlled by metabolic states, and at least some metabolic processes when accelerated modify the physiological conditions in such a way as to promote growth. Several workers have shown that a vigourous respiration is necessary for growth to occur. Bonner found with coleoptile fragments that extension did not occur in the absence of oxygen or in the presence of cyanide.

Brown & Sutcliffe have shown with root fragments that growth is limited when the partial pressure of oxygen is reduced below that of air. The promoting effect of a vigourous respiration is, however, a complex phenomenon, and in view of results considered below it is probable that it involves two aspects. It is probable that it is a necessary condition for the promotion of certain other processes on which growth immediately depends, and that certain of the phases of respiration are directly linked to growth.

That these two aspects are involved is suggested by a consideration of the effects of sugar and potassium on growth. Thimann & Schneider reported that the growth of coleoptile fragments is enhanced by the provision of sugar, and Bentley has recently made a similar observation. Brown & Sutcliffe reported that in the absence of sugar the growth of root fragments is limited and that vigourous growth is only given in the presence of this nutrient. These workers analysed the sugar effect in terms of absorption and accumulation of sugar, synthesis of cellulose and respiration.

They found that all these processes tend to increase with increasing concentration of sugar, but only when the cultures are exposed to air. In low partial pressures of oxygen they found a low respiration, a low absorption, no accumulation, and no cellulose synthesis. It was suggested that a high partial pressure of oxygen promotes a vigourous respiration which stimulates absorption. The rapid absorption in turn promotes the accumulation of sugar which establishes the osmotic gradient along which the absorption of water can occur. The rapid sugar absorption also provides the conditions for the synthesis of cellulose and thus stimulates the expansion of the wall.

This interpretation is consistent with the observation that there is no quantitative relation between respiration and growth. In this context respiration provides the necessary condition for the promotion of sugar accumulation and cellulose synthesis and is not directly linked to them. At the same time these are energy consuming processes, and a specific mechanism is presumably involved in effecting the energy transfer.

Bonner finds with coleoptile fragments that sodium arsenate inhibits growth but has little effect on respiration, and he suggests that this is due to the inhibition of the transfer of energy-rich phosphate bonds generated in respiration to reactions immediately involved in growth. In the present connexion it may be suggested that respiration provides a source of energy-rich compounds which become involved in reactions immediately connected with sugar accumulation and cellulose synthesis.

The effect of potassium suggests that there may also be a direct link between respiration and growth. The influence of potassium in the growth of coleoptile tissue is uncertain; Thimann & Schneider found a stimulation, Frey-Wyssling an inhibition, and Bentley an inhibition. Whatever the position may be with coleoptile tissue, with root tissue it is clear and unequivocal.

Brown & Sutcliffe found that in the presence of potassium growth and respiration are enhanced while cellulose synthesis and the absorption of sugar are not. The enhanced growth cannot be attributed to a greater synthesis of cellulose. Also since the absorption of sugar is not increased when growth and consequent water absorption are raised, it follows that the enhanced growth occurs with a lower internal concentration of sugar.

Brown & Sutcliffe suggest this might be due either to the adjustment of the internal osmotic pressure through an accumulation of potassium or to the stimulation of a secretion mechanism such as that proposed by Bennet-Clark, Greenwood & Barker. It is significant, however, that potassium has no effect either on growth or respiration in the absence of sugar. Cartwright has recently examined the internal potassium concentration of the fragments used by Brown & Sutcliffe in the presence and absence of sugar and finds there is little difference.

This suggests that the internal potassium concentration as such can have little effect on the osmotic gradient, and that the accelerating effect of potassium on growth is due to the stimulation of a water-secreting mechanism acting through a primary stimulation of some phase of the respiratory cycle. The interpretation proposed here would involve an intimate connexion between some limited phase of respiration and the secretion of water in the sense that an inhibition of this phase alone would arrest the secretion of water.

Commoner & Thimann have provided evidence for an intimate link of this general type between growth in the coleoptile and respiration. They find that 10-5M-iodoacetate arrests growth but only depresses respiration by about 10%, and they suggest that growth is controlled by this fraction of the total respiration to which it is linked.

A vigourous respiration may also be involved in another connexion. During growth in the intact organ there is a net increase in protein content. Even if the protein content does not increase, however, it is probable that protein synthesis must occur, and that this synthesis is necessary for the promotion of growth. The synthesis is an energy-requiring process, and can only occur when respiration is vigourous.

The growth may be enhanced by the stimulation of certain metabolic processes. It is clear, however, from curves, that the overall growth process cannot be analysed in terms of such processes alone. The effects of different concentrations of the same nutrient on the final lengths of the fragments suggest that growth is also determined by a complementary set of processes the stimulation of which tends to arrest it.

The final length in 0.25 % sucrose is lower than it is in 2 %. The lengths do not converge to the same limiting value at different rates depending on the external concentration. A similar situation is shown by the data of Schneider on the effect of different concentrations of sugar on the growth of coleoptile fragments.

If the stimulation of growth depended on the stimulation of a particular set of metabolic processes, with isolated fragments growth might be expected to continue until one or a complex of conditions became limiting. The limiting condition might be such as the protein level, the distance of the nucleus from the two ends of the cell, or some property involved in the geometry of the system.

Such a limiting condition, however, would be common to all members of an experimental series, and with different levels of a single nutrient they might all be expected to reach the same final length, but at rates which are directly related to the external nutrient concentration. Clearly some other factor is involved in the process, and this it may be suggested is the operation of an arresting mechanism which stops growth before a hypothetical maximum length can be reached in at least some series.

Other evidence is available which supports this conclusion. It has been shown that in the coleoptile the same concentration of heteroauxin is less effective if added 12 hr. after excision than it is if applied immediately. Similarly with root fragments, if these are transferred to water for 6 hr. and then to a sugar solution, the final length is less than it is if the sugar is supplied immediately. Clearly in the interval during which the stimulant has not been available changes have occurred which tend to restrict subsequent growth.

Secondly, there is evidence that with certain low partial pressures of oxygen the capacity for growth is retained longer than it is in air. When the oxygen level is very low the conditions of anaerobiosis have the normal effect of damaging the tissue. In 10 % oxygen, however, the arresting mechanism tends to be inhibited. In a series of experiments conducted by Sutcliffe fragments of maize roots were cultured in air and in an atmosphere containing 10 % oxygen, and at intervals during the experimental period cultures were transferred from the lower to the higher partial pressure.

With regard to continuous exposure to air and to an atmosphere containing 10 % oxygen, it is evident that while the rate of growth is lower in the lower partial pressure the time during which growth proceeds is very much greater. Clearly in the lower partial pressure the processes which promote growth are depressed giving a lower rate, while those tending to arrest growth are inhibited giving a longer period of growth.

The effects of transferring from the lower to the higher partial pressure are consistent with this interpretation. After transference the rate is accelerated, but this occurs after a period during which, although some growth has occurred, the growth-arresting processes have been inhibited. The net result is that the final lengths of the fragments are considerably greater than they would have been if they had been continuously in air.

Thirdly, it has been shown with root fragments that before growth ceases there may be a decrease in respiration. In intact roots it is probable that there are decreases in protein, the activities of certain enzymes, and respiration, either before or at about the time when growth ceases. The evidence suggests that progressive changes continue during growth, which when they have reached a certain critical level depress the rates of various metabolic processes, and it is as a result of such decreases that growth ceases.

The data obtained with isolated fragments which suggest the operation of an arresting mechanism do not indicate the nature of that mechanism. It may be suggested tentatively, however, that this involves the progressive inactivation of an enzyme system, in which case growth may be interpreted in terms of the rate per unit quantity of an enzyme or enzyme complex and the simultaneous absolute or relative reduction in the total quantity of this complex during growth.

In this situation it may be taken that the rate of the reaction per unit quantity of enzyme remains constant. The curve represents the change in the content of enzyme with time. Clearly the distance from this curve to the abscissa represents the rate of growth, and the area under the curve the total amount of growth made.

It is immediately evident that at any one time the growth rate is higher, the total amount of growth made is greater and growth proceeds for a longer time. The lengths of the vertical lines here represent the rate per unit quantity of enzyme.

It is evident that when the total amount and the rate of growth are affected by the rate per unit quantity of enzyme and by the rate of reduction of the total quantity, then the only factor that can affect the time of growth is the rate of reduction of the quantity of enzyme, and if the time of growth is not affected then no effect on the rate of inactivation can be involved.

Thus it may be suggested that since potassium increases rate and final length but has no effect on time this is due to a stimulation of the promoting mechanism. On the other hand, sugar which undoubtedly stimulates the promoting mechanism, since it increases the time of growth, probably also inhibits the arresting mechanism.

It may be emphasized, however, that the present interpretation suggests that an enhanced growth in terms of rate and time could occur by a simple inhibition of the arresting mechanism only. It has been assumed that the rate of inactivation increases with time. It is evident, however, that the general character of the interpretation is not affected by the form of the inactivation curve.

The arresting mechanism has been considered above in terms of a reduction in a total quantity of enzyme involved in growth. This is suggested by the changes in certain metabolic processes (notably respiration) that occur during the course of growth. Further, measurements have been made of the changes in the activities of certain enzymes, and it has been found with root fragments in certain instances that there may be a decrease in invertase activity and an increase in acid phosphatase activity (unpublished data). Thus changes during growth do occur in the activities of particular enzymes.

These changes may be due to the production of specific inhibitors or to a transformation of one enzyme into another, in either case leading to a reduction in the total quantity of active enzyme. Another possibility may, however, also be considered. It is that growth arrest may be due to a relative and not an absolute decrease of a particular enzyme complex.

If two enzymes increase that compete for the same substrate the reaction catalysed by one enzyme might be virtually suppressed by a very much greater relative increase in the second. It is possible that the inactivation is of this type in the intact root, where there is a considerable increase in total protein during growth. On the other hand, an absolute decrease may be involved in the arrest of the growth of the fragments.

It was suggested above that the growth of fragments must be determined by four primary physiological conditions which in turn are regulated by metabolic states. It is clear, however, that the states that arrest growth must operate through these conditions no less than do those which promote it. The rate of growth of the root fragments used by Brown & Sutcliffe decreases with time. This decrease, is an expression of the increasing effect of the arresting mechanism, and should therefore be reflected in a corresponding change in one or more of the primary physiological conditions.

Brown & Sutcliffe, however, found that during growth of the fragments there are no changes in the rates of sugar absorption, sugar accumulation and cellulose synthesis. It is probable, therefore, that the effect of the arresting mechanism is expressed either through a change in the physical state of the wall or in the water-secretion mechanism, the effect of which might of course be determined by a change in the permeability of the protoplast to water.

Chapter 7

Transpiration

The transpiration of a plant may be defined as the evaporation of water from its aerial parts, such as the stem, twigs and leaves. The amount of water lost by this method is in many cases considerable. The loss of water from cotton crop by transpiration is about 50 tons per acre per day or 3 pints per plant.

The main mass of transpiration takes place from the leaves, but it has also been found to occur in the stem as well as in the twigs. It is well known that transpiration may be divided into two classes: *cuticular* and *slomatal.* Cuticular transpiration is the amount of water lost through the cuticleofa plant ; stomatal transpiration is the loss of water 'through tne stomata. In ordinary mesophytic plants stomatal transpiration represents 80 to 97 per cent, of the total water-loss, whereas cuticular transpiration only represents about 2 to 3 per cent, of the total loss. It has been found that even the very thickest cuticles transpire to a certain extent. In some leaves the cuticular and stomatal transpiration is approximately equal.

A good example of this is to be found in the Jamaican rain forests, where the cuticular transpiration represents 58 per cent, of the total transpiration of the leaves. The leaves possess the power of cutting down their relative transpiration during the months in which aridity is increasing. The plant possesses two distinct types of leaves.

A mesophytic type of leaf is present during the cool months of the year. This is shed when arid conditions arrive, and is in turn succeeded by a xero-phytic form. Curiously enough, it was ascertained that when discs of uniform section were cut from the two types of leaves, those obtained from the xerophytic leaves lost water 1-44 times more rapidly than discs cut from the mesophytic form; although if the water-loss were calculated on a dry weight basis the discs from the xerophytic leaves only lost water 0-78 times as fast as the mesophytic discs.

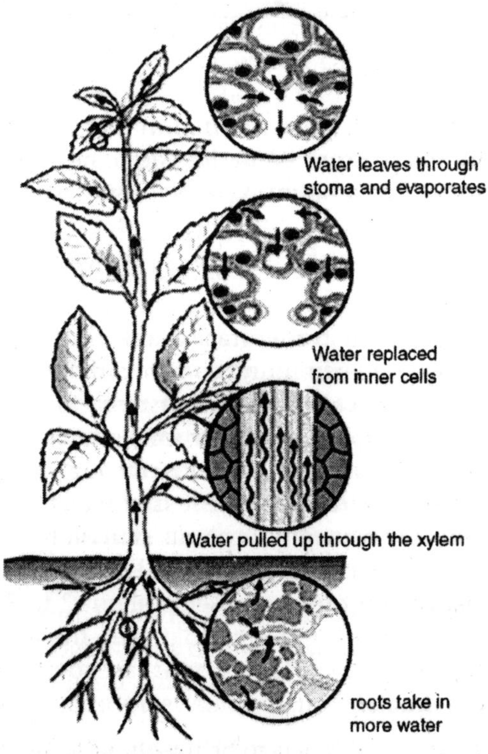

Fig. Transpiration

On the other hand, Shreve discovered that while arid conditions were in existence the xerophytic leaves secrete a viscous brown fluid which appears to play a considerable part in hindering excess of transpiration. In stomatal transpiration the diffusion of water vapour takes place through the open pores of the stomata. The physical side of the matter has been very fully investigated through the now classical investigations of Brown and Escombe, Rcnner, Jeffries and others, and will not, therefore, be considered here.

Since however, the main mass of the transpiration of a plant takes place through the stomata, it is obvious that external factors play a very considerable part in determining the rate of water-loss. Stomatal diffusion is a process of evaporation. Evaporation from the wet cell walls of the mesophyll cells takes place, and is markedly affected by the external conditions of the atmosphere prevailing at the time. Energy is expended in the process, of the incident sunlight falling on the leaf of *Polygonum Weyrichii* was used in transpiration, and only some 1 per cent, or less utilised for photosynthesis.

FACTORS AFFECTING THE TRANSPIRATION RATE

The factors affecting transpiration may be divided into two classes: external and internal.

External Factors.

1. Humidity of the atmosphere.
2. Wind.
3. Temperature.
4. Barometric Pressure.
5. Light.

Internal Factors.

1. Stomata.
2. The water-content of the mesophyll tissue.

THE HUMIDITY OF THE ATMOSPHERE

If the atmosphere surrounding the leaf be saturated with water-vapour, no gradient of water concentration can be set up between the intercellular spaces of the leaf and the external atmosphere. In such circumstances no evaporation can take place, and, in consequence, transpiration will be greatly diminished.

The main concern in such a question as this, is the water deficit of the atmosphere and the temperature must be known for a knowledge of this factor. Any rise in temperature will alter the water vapour pressure and cause an increase in the water deficit. Similarly, a decrease in temperature will cause a decrease in the water deficit. If transpiration be nothing more than a process of evaporation from the wet walls of the mesophyll tissue, there should exist a simple relation between the moisture present in the atmosphere and the transpiration rate.

The first attempt to ascertain for *Hedera helix* the relationship between the transpiration rate and the humidity of the atmosphere when the temperature was kept constant. He discovered that with decrease in humidity there was increase in the transpiration rate, and that a straight line graph was obtained, *i.e.*, there was a simple linear relation between the two. The line, however, when produced backwards, did not cut the axis at the 100 per cent, saturation point, but at the 105 per cent, saturation point.

The explanation of this curious result proved simple. The living plant respires in the course of its metabolic activities, and the temperature in the immediate vicinity of the mesophyll cells will on this account be higher than that of the surrounding atmosphere. Hence the air in the direct neighbourhood of the mesophyll cells will not be saturated when the external atmosphere is in that condition. It is on this account that the value 105 per cent, was obtained instead of the expected 100 per cent.

From the results of these experiments, Darwin was able to calculate the temperature of the mesophyll compared with the surrounding air. He worked at 16° C. and the vapour pressure at this temperature is 13-51. This value, plus 5 per cent., gives a vapour pressure of 14-2. The corresponding temperature to this vapour pressure is 16-8° C. Thus the mesophyll cells were 0-8° C. higher in temperature than the surrounding air.

These observations of Darwin have been in large measure confirmed by Henderson using more controlled conditions. He found that in *Hedera helix* his curve did not cut the axis as far back as Darwin's, and that the temperature of the mesophyll cells was 0-4° C. and not 0-8° C. higher than the surrounding atmosphere.

Wind

The effect of wind on the transpiration rate is to remove the layers of saturated air or partially saturated air which cover the leaf surface. This will lead to an increase in the transpiration rate.

Temperature

In connection with transpiration, temperature has also to be taken into account. Temperature plays no direct part in affecting the water-loss from leaves, but acts in the indirect capacity of affecting the saturation deficit of the atmosphere. Increase of temperature will lead to an increase of the saturation deficit and thus to an increase in the rate of transpiration.

Humidity, temperature and wind can all be expressed in terms of the evaporating power of the air, and it is a comparatively simple matter to determine at what rate water is evaporated from a surface.

Light

Light materially affects the transpiration rate. When, however, the effect of light is being investigated, it is of the utmost importance that all the other factors involved are kept strictly constant. The matter becomes difficult when investigations areconducted under field conditions in the open.

It is therefore more convenient for comparison to use a record of the rate of evaporation from a standard water surface at the same time. This expression summarises the influence of the prevailing atmospheric conditions and makes it an easy matter to detect any differences between the effects of these on transpiration rate and evaporation rate.

Some years ago, a method of measuring the evaporating power of the atmosphere, a value which he termed E. If T represented the transpiration rate, the ratio T/E, referred in both cases to unit area, was termed by him the " relative transpiration." Relative transpiration eliminates the direct evaporating power of the air and the physiological behaviour of the leaf is exhibited.

The main difficulty in this connection is to measure E in a satisfactory manner. It would, of course, be possible to place an uncovered basin in the open and find the loss of water by weighing at intervals. This method, however, has several disadvantages in actual practice. One of the main sources of error in such a procedure is this: the rate of evaporation will alter as the water level falls below the rim of the vessel. Livingston escaped from the difficulty by means of a device known as the " porous cup-atmometer." Various types of these are now in use. A common variety consists of a flask fitted with a cork and glass tube at the other end of which is attached a porous porcelain candle.

Flask, tube and porous candle are filled with water, and the evaporation determined by weighing at intervals. Another method of comparing T/E is by means of Pische's device. This, in essence, is very simple. It consists of a piece of filter-paper placed over the mouth of a tube containing water. It is not, however, very convenient to use in the field.

The question arises as to how far the atmometer responds in the same way as the leaf. Livingston assumed that by his conception of relative transpiration the direct effect of the atmospheric conditions on the transpiration rate could be neglected in the interpretation of his results. The main assumption underlying this principle is, that changes in the atmospheric environment affect equally the rate of transpiration of a plant and the rate of evaporation from a water surface.

As Knight has pointed out, the assumption is scarcely warranted by the facts. Livingston and Briggs and Shantz have shown that atmometers of different sizes and shapes are not strictly comparable under changing conditions. A change in environment which doubles the rate of evaporation from one atmometer does not necessarily double the rate from another of different size or shape. In exactly the same way the size or shape of the leaf may modify the influence of changes of external conditions on its transpiration rate.

It appears, therefore, that it is not possible to compare one atmometer with another, or one leaf with another under changing external conditions, so that it seems scarcely justifiable to compare an atmometer with a plant.

A further criticism advanced by Knight is to be found in the structure of the leaf. A portion of the path along which the diffusion stream passes from the evaporating mesophyll cells to the outside air, is situated actually within the leaf and, being protected by the epidermis, is not, therefore, subject to the direct influence of movements of the outside air.

In the case of the atmometer, however, the whole of the diffusion stream is exposed to air movements, and it is theoretically possible by means of a sufficiently rapid current of air to reduce the moisture in the air close to the evaporating surface to the same concentration as that in the general atmosphere. In the case of the leaf, it is only at the surface of the epidermis and not of the evaporating mesophyll that this minimal concentration can be obtained, so that it is to be expected that changes in the speed of air movement will have less influence on the transpiration rate of a leaf than on the rate of water-loss from an atmometer.

From a large series of experiments, Knight has shown that changes in the velocity of wind affect the plant and atmometer in different ways, and it is only when the air movements are maintained at a constant value that this method of comparison by use of relative transpiration can be used to eliminate changes in relative humidity and temperature on transpiration ; since it has been found that these two factors act equally upon transpirationfrom a plant and evaporation from an atmometer.

It is clear then, that *only when wind velocity is constant* does relative transpiration give a satisfactory measure of the intrinsic transpiring power of a plant. A number of years ago Livingston compared the transpiration of the succulent *Euphorbia capitellata* with the evaporating power of the air, and further ascertained the relative transpiration (T/E) at the same time. An examination of his curves show a general similarity, but with an important difference.

The highest point of transpiration was found to occur between 10 a.m. and 12 noon and this increase was followed by a steady decrease. The evaporation rate, on the other hand, increased between 2 p.m. and 4 p.m., at which point the maximum was reached. The prevailing external conditions are evidently of such a nature as to cause maximum evaporation in the late afternoon, whereas some reaction in the plant has led to a diminution of water-loss at an earlier time. The same result is reflected in the graph of relative transpiration.

If both the transpiration and the evaporation had been affected in a similar way, this latter curve should have been a straight line. The actual result obtained shows that transpiration is increasing more rapidly than evaporation till the maximum is reached, and the subsequent increase in the evaporating power of the air is not reflected in the transpiration.

A further point in this connection is that transpiration increases greatly in light, and falls in the dark. The light produces some considerable change in the leaf, and, in actual fact, brings about an increase ordecrease in the size of the stomatal openings through which the main mass of transpiration is taking place.

STOMATAL REGULATION

The question as to whether stomatal aperture does or does riot control the transpiration rate has led to a good deal of controversy which has only recently been settled. Lloyd, at the end of his elaborate experiments on the relation between stomatal opening and transpiration rate, came to the conclusion that there was no correlation between them. Trelease and Livingston arrived at a similar coincident increase of osmotic pressure.

It was shown by Leitgeb, however, that the stomata open in the light in the absence of carbon dioxide, while the later experiments of Lloyd, Iljin and Loftfield, showed that the opening in light is accompanied by a decrease in the starch - content of the guard-cells, and at closure the starch-content is increased. In other words, the behaviour of the guard-cells was the direct opposite to that of the assimilating cells of the mesophyll. Such being the case, the decrease in the starch-content must lead to a considerable increase in the osmotic pressure of the guard-cells giving the necessary conditions for stomatal opening.

Lloyd, Loftfield and others, have supposed that the change of starch into sugar and sugar into starch is in the nature of a reversible enzymic reaction. The question arises as to how far does the influence of light affect the direction of the equilibrium. Leitgeb, F. Darwin, Lloyd, have shown that a reduction in the partial pressure of the carbon dioxide surrounding the leaf led to the opening of the stomata; while Lindbauer discovered that increase in the partial pressure of the gas led to the opposite result.

Weber connected these apparently contradictory facts with the action of the light, and assumed that the concentration of carbon dioxide in the guard-cells was regulated by the rate of its photosynthetic absorption. Both Lindbauer and Weber have shown that remarkable physiological reactions accompany the functioning stomata. The plastids changed their position, the nucleus underwent changes in size and shape and the open and closed stomata showed differences in their power of reducing silver nitrate.

More recently Sayre and Scarth have shown that these various changes observed in the guard-cells are secondary phenomena of a simple common case. Using solutions which readily penetrated the guard-cells, such as solutions of ammonia and acetic acid, Scarth found that the stomata opened in both alkali and acid, but more fully in the former.

In an intermediate range of *pH* (5-5 to 7-0) for *Zebrina pendula,* the stomata remained closed, but opened in increasing concentrations of both alkali and acid up to the limits of injury. It was further discovered that starch made its appearance in the *pH* zone of closure and disappeared in that of opening, and these changes were reversible. Scarth also ascertained that the sap of the guard-cells under normal conditions possessed a higher *pH.* when the stomata were open, and that a low jpH was registered when they were closed. The range appeared to stretch from *pH* 7 to 4-5.

Sayre obtained very similar results for *Rumex patientia.* The guard-cells showed a more alkaline reaction when the stomata were open, and the movement could be controlled at will by variations in the *pH* of the guard-cells. The stomata opened within the range of *pH* 4-2 to 4-4 and closed on both sides of this. Sayre considered that there was an optimum *pU.* for the enzymic hydrolysis of starch, whereas Scarth thought from the fact that there are two widely separated optima, and that the direction and rate of the reaction appear to have no relation to the relative concentrations of the reagents, that stomatal opening is not explicable on a simple enzymic basis.

The metabolism of the stomata of white-margined *Pelargonium* leaves has been investigated by Klimmler, who showed that they contained abundant starch—more, in fact, than the guard-cells of normal green leaves—and that they opened in light and closed in the dark. Very wide opening only occurred if the water supply were highly favourable. These results appear to show that chlorophyll is not an essential part of the plastids of the guard-cells, and it is possible that they are in the nature of leucoplasts.

External conditions, such as humidity, temperature, etc., all play a part in influencing stomatal opening. Loftfield has shown that temperature markedly affects stomatal movement and that for a rise of 10° C. the rate of movement is approximately doubled. Excessive transpiration may lead to a loss of turgor by all the cells of the leaf, including the guard-cells, when collapse and closure of the stomata takes place.

The investigations of Knight have largely cleared away the conflicting statements that have been made from time to time concerning the effect of wilting on stomatal behaviour. F. Darwin considered that the stomata opened widely in the early stages of wilting, and Knight has confirmed this observation. When the leaf is losing water more rapidly than water is supplied to it, there is a temporary increase in stomatal opening. Only when the wilting is very pronounced does closure take place.

The complexity of the stomatal mechanism is shown by the recent work of Maskell, who discovered a well-marked rhythm in the stomata of *Primus laurocerasus.* Under constant artificial light intensity the stomata began to close towards the evening and to open about midnight.

It must be confessed that our knowledge of the underlying mechanism which controls and regulates stomatal behaviour still leaves much to be discovered.

THE EFFECT OF LIGHT ON TRANSPIRATION

Although a great deal has been done in the way of discovering the effect of stomata on the rate of transpiration and also a number of facts and figures arc available for the transpiration of the leaf as a whole, very few attempts have been made to ascertain the direct effect of changing external conditions on the mesophyll alone, although it is from these cells that the main water loss occurs.

The action of the mesophyll lias in most cases never been separated from that of the stomata. The difficulty is to remove the action of the stomata. It was suggested by Knight that leaves from which the epidermis had been stripped might be employed, but the method is drastic, and there is the further danger of the drying of the mesophyll cells except in atmospheres of high humidity.

The other method was that employed by F. Darwin. The essentials of this method are to exclude the action of the stomata by stopping them up with some substance like vaseline and then placing the intercellular spaces in contact with the atmosphere by means of slits in the leaf tissue. Such slits were usually made with a scalpel between the principal veins of the leaves so as to interfere as little as possible with the water supply. Using this method,

Darwin attempted to determine the effects of changes of light and humidity of the atmosphere on the mesophyll tissues of the leaf. He came to the conclusion that the leaf increased its transpiration by amounts that varied between 10 to 100 per cent, of that in the dark. An increase of the latter order is certainly most remarkably high. To bring about such a doubling of the evaporation rate from a moist surface in air of constant evaporating power would require an increase of temperature of at least 10° C. From a careful consideration of Darwin's data, Henderson was able to ascertain two possible sources of error.

Firstly, Darwin's observations were usually begun immediately after smearing the leaves with vaseline and slitting, thus making no allowance for " wound-shock response," and secondly, external conditions were seemingly quite uncontrolled and largely unknown. The plants were in some cases removed from the window of the laboratory into the dark-room, apparently without accurate observations being made of humidity and temperature.

Henderson, by means of a highly ingenious piece of apparatus, has repeated and extended Darwin's observations. The plants used in these experiments were *Hedera helix* (Darwin's original plant), *Eupatorium adenophorum* and *Aster.* It was found that the transpiration at first rose to a high value after slitting,, and then settled down to a steady rate in the course of an hour or so.

The cause of this strange behaviour was not ascertained. It was during this preliminary unsettled period that all Darwin's readings were taken. Since all his experiments were begun in the light and the plants removed to the dark-room when the maximum was reached, it was not surprising that a drop was shown in the transpiration rate in the dark, and it is also of significance in this connection that when the plants were again placed in the light, the second curve nowhere reached the same value as the first. Had Darwin started his experiments in the dark, then in all probability his results would have been reversed.

It was shown by Henderson at the conclusion of his experimental work that light does have a small direct effect upontranspiration from the mesophyll; the order of increase being from 4 to 5 per cent, of that in the dark. The exact action of the light on the mesophyll is unknown. Henderson advanced the view that it may be due to the presence of unknown biological complexes, and is possibly in the nature of a photo-chemical reaction.

It cannot, however, be said that the explanation is of a very helpful nature. In any case this action of the light shows that transpiration is not solely a simple process of evaporation. It is possible that a secretion of water from the mesophyll cells of the leaf takes place; the cells actively secreting water or some solution on their outer surfaces. Here again, it must be confessed that the word " secretion " in reality only acts as a cloak for our ignorance of the chemical and physical phenomena underlying the whole process.

IMPORTANCE OF TRANSPIRATION TO PLANT

Transpiration exists in all land plants. Whether it be an advantage or a disadvantage, the fact remains that it is quite unavoidable. The absorption of carbon dioxide from the air by the leaf is a prime necessity for the synthesis of carbohydrates by the living plant, and it follows that there must be some suitable mechanism of gaseous exchange in the organisation of the plant body, and this exchange is principally performed by the stomata of the leaf.

As a corollary, it follows that if stomata are present, water losses must occur. The aerial portion of the plant soma is covered with cuticle, and were this cuticle to extend over the whole surface of the plant, it would be unable to carry out its normal function of carbon dioxide absorption and, of necessity, its metabolic activities would cease.

In the case of ordinary mesophytes the temperature of the leaves is not much higher than that of the surrounding atmosphere, usually about a degree or so. In high winds the temperature may even fall below that of the air. Even when leaves are cut the temperature only rises one or two degrees from " wound-shock." It must also be remembered that in the case of the ordinary type of dorsiventral leaf with its flat surface, there is a very large area exposed to the air which would tend to prevent any rise in temperature should external conditions of drought and heat arise.

The most exact work on this subject has been carried out by Clum using leaves of *Fuchsia speciosa, Brassica oleracea* and *Syringa vulgaris*. The temperature of the leaves was determined by means of thermo-couples under a number of experimental conditions. The leaves were always found to be at a higher temperature than that of the surrounding air. The effect of shading was found to cause a sudden drop in the temperature ; further the direct effect of sunlight was very much marked, and the particular angle at which the rays struck the leaf had an effect on its temperature. In no case, however, could he find any correlation between leaf temperature and transpiration.

A second claim that has been put forward to show thatjrans-piration is an advantage to the plant is that it rids it of excess water and allows of the rapid absorption of salts. A considerable amount of evidence was at one time forthcoming in support of this view, but it has since been shown to be erroneous. The process of transpiration has nothing to do with the absorption of salts ; the two processes work independently of one another.

The absorption of salts is a question of the salt and water equilibrium of the cell, and is not connected with the evaporation of water from the leaves. Moreover, the concentration of salts in the plant is very much higher than the concentration of the corresponding salts in the soil. An early experiment of Ilassel-bring bears on this problem. He grew tobacco seedlings in the open where transpiration would be normal, and a second series under calico to reduce transpiration. At the end of a given time the ash-content of the two sets was determined, and was found to be identical in both series.

The shade plants had absorbed 35 litres of water and the sun plants 46 litres or 30 per cent. more. The amount of mineral matter absorbed by the plants was therefore not proportional to the transpiration, nor did the plants with the lower transpiration suffer with regard to salt absorption. Later experiments by other investigators have given substantially the same results. Muenscher grew barley in water culture, and, by suitable arrangements, was able to cut down the transpiration rate.

He ascertained that there was no difference in the ash-content of the series grown under conditions which reduced transpiration; the ash-content could only be correlated with the growth of the plants. Mendiola found that in tobacco a reduction in the amount of transpiration led to an increase in the dry-weight and a decrease in the ash-content. On the other hand, his results support the view that there is no proportionality between the amount of transpiration and the amount of salt absorbed by the plant.

It is quite possible, however, that transpiration might increase the rate of transport of salts from one part of the plant to another, once the salts had effected an entrance. The rate of the transpiration stream is rapid, while the passage of salts from cell to cell is slow. It has been shown that salts are present in the xylem, and it may well be that the transpiration stream conveys salts from the lower to the upper regions of the plant. There may be a critical rate of the transpiration stream below which the water fails to carry the salts in the xylem with any rapidity.

THE TRANSPIRATION CURRENT

The question of how water is conveyed from the roots to the leaves of plants has long exercised the minds of plant physiologists, and a number of theories have been advanced to account for the process. In some plants, such as the Blue Gums of Australia, which are often as much as 300 feet high, the question of water transport becomes a serious one.

It has been conclusively shown that the path of the current is through the xylem. It is easy to demonstrate this fact by simple experiments. If a cut shoot be placed in a weak solution of such a dye as eosin, the dye will only be found to be present in the trachea'. Again, a ringed tree will continue to transpire for a considerable time, while water plants which transpire but little, have only weakly developed woody tissues.

There is need here of a dynamic and not a static explanation of the process ; in other words, an explanation that will adequately account for the continuous drive of water up the cavities of the xylem. Such phenomena as evaporation, capillarity, atmospheric pressure and root pressure have all been invoked to account for the ascent of sap. Capillarity and atmospheric pressure are both, static processes; after water has risen a certain distance, no further rise can occur. Root pressure is a dynamic process, but it, however, is low in summer, when transpiration is at its maximum, so that root pressure will not adequately account for the ascent of sap in plants.

It is, therefore, not surprising that the alder investigators fell back on vital theories to account for the rise of the water current, and assumed that some kind of pumping action was initiated by the living cells in the xylem. such as the xlyem parenchyma. Strasburger's classical experiments conducted many years ago, showed quite conclusively that water would still ascend in plant stems when the living cells had been killed, and thereby sounded the death knell to any vital theories, and physiologists were forced to fall back on purely physical explanations.

THE COHESION THEORY OFF THE ASCENT OF SAP

Theory originally propounded by Dixon and Joly in the 'nineties, that the ascent of sap was due to the cohesion of the water columns in the cavities of the xylem and was drawn off from above by the leaf, has been very fully dealt with by

Dixoifln his monographs—*Transpiration and the Ascent of Sap,* and the *Transpiration Stream.*

The theory Dixon has put forward and supported by a considerable amount of quantitative and qualitative data, has rather tended to overshadow all other work on the matter. The chief disadvantage of the theory is that it explains too much.

Briefly stated, the Dixon and Joly cohesion accounts for the ascent of sap by taking into consideration thte fact that water possesses strong cohesive properties and that quite considerable pressures may be developed in this way, in fact, pressures as high as 300 atmospheres have been recorded. According to Dixon, then, the ascent of sap takes place as follows: the water in the conducting tracheae hangs in fine columns cohering to the walls of the tracheae. From this initial condition we can proceed to a consideration of the passage of water from the soil through root and stem to the leaves.

Through the suction pressure of the root-hairs and cortical tissue, water is removed from the soil and passes into the xylem. By means of its cohesive properties the water adheres to the wall of the cavities of the xylem in fine columns, and from these columns, which are in a high state of tensile strain, it is withdrawn by the mesophyll tissues of the leaf. These hanging columns of water can exert a pull equal to that of a fine steel wire, but they can only do this while they are attached to the walls of the containing tracheae. The initial force in the process is exerted by the root.

Although the cohesion theory of the ascent of sap explains in a remarkable manner a large number of facts that have been observed about the transpiration stream, in actual practice it is difficult to devise crucial experiments which would definitely settle the truth or otherwise of the theory. From the Dixon point of view the matter is reversible, that is to say, the plant should react equally well if it were placed upside down. It must be remembered however, that it would be almost impossible to obtain the same contact between the particles of the soil and the leaves as exists between the root-hairs and the soil and the leaves as exists between the root-hairs and the soil.

The main criticism that has been levelled at the theory is the presence of air-bubbles in t,he wood. The presence of air-bubbles immediately destroys the stability of the hanging columns and would cause their collapse. It is impossible to state definitely whether the air-bubbles seen in cut sections are not brought there by mechanical manipulation, or whether their presence is natural in the wood of the living plant. The investigations of Bode seem to indicate that there are no bubbles in the xylcm normally, and that even in the last stages of wilting the water-content of the tracheae is continuous.

Dixon accepted the presence of bubbles in the xylem. The difficulty is to remove them. The suggestion has been made that perhaps root pressure in the spring is sufficiently strong to get rid of them. But when the whole plant is transpiring rapidly, a negative pressure is set up in the wood, and the bubbles are then under a negative pressure ; when a positive pressure is introduced they will suffer compression.

A further difficulty that arises from the Dixon point of view is the fact that tracheids should be more efficient than vessels for the conductipn of water. In Dixon's own words: " The salient feature of the wood is the subdivision of the waterways by an immense amount of transverse and longitudinal divisions into minute compartments the vessels and tracheids. For the system the function of which is the conducting of water this is evidently the most unexpected configuration. It is true that the partitions are permeable to water; but when a considerable distance has to be traversed the sum of the distances opposed by the walls becomes very important.

It is evident from the persistence of walls in the development of the water-conduits of a plant, introducing as they do an immense resistance to flow, is inexplicable on any view that regards the water being forced through the stem by some unknown force. The cohesion theory, on the other hand, gives a ready explanation, for it confers stability on the tensilely stressed transpiration/stream. From this statement it is evident that tracheids with their numerous short compartments would give greater stability to a tensilely stressed stream of water than vessels in which the cross-partitions are relatively far apart.

Yet, such being the case, it is surely a curious anomaly that our present dominant flora, the Angiosperms, have practically without exception adopted vessels in place of tracheids. The various objections raised in this discussion to the unqualified acceptance of the colics ion theory, as Professor V. H. Blackmail once remarked: " Somewhat dilute the milk of the cohesion theory." It must be admitted that at the present time there is no adequate explanation of the ascent of sap, and although the cohesion of water possibly plays a considerable part in the matter, it is probably only a portion of the whole truth.

In spite of the work of Strasburger and others, the vital theory of the ascent of sap has apparently still a hold on some minds. At the present time the most ardent exponent of such a theory is the Indian investigator^Sir Jagardis Chuiulra Bose. According to this worker, water is driven through the plant by some kind of pulsatory activity of the living cells, these cells beyig identified with the innermost cells of the cortex. The methods of reasoning which lead him to make these assertions arc remarkable and peculiar.

Bose reviews the various theories that have been put forward from time to time to account for the rise of sap in plants and any view that the rise depends on purely physical grounds is considered to be quite untenable. The work of Dixon and Joly is summarily dismissed in three or four lines, Strasburger is held to have misinterpreted his results, and Bose then quotes a statement from Pfeffer's text-book of plant physiology written over forty years ago wherein it is stated:—

"How and by what means the water is so rapidly transmitted even to the summit of the tallest trees has not yet been satisfactorily explained. It has unfortunately not yet been determined whether the aid of living cells is quite necessary."

Apparently all that has since been discovered is of no consequence to Bose, who sets forth his views backed by a series of experiments chiefly interesting for the highly complex apparatus that was employed.

According to Bose, a pulsatory activity is initiated in the roots, affecting the absorption of water from the soil. A similar activity pumps water up through the stem from cell to cell and also injects water into the vascular tissue of the xylem along which it is physically transferred. The water conveyed by physiological conduction and physical convection reaches the leaf, where it is distributed along the veins and their numerous ramifications, and is eventually excreted and transpired into the intercellular spaces, whence it finds its way to the atmosphere outside by evaporation.

To establish this standpoint, Bose has taken the following experimental line. Similarly, by means of his " crescograph," Bose asserted that growing organs also exhibit an autonomous activity of a pulsatory nature. Having convinced himself of the truth of these two statements, Bose further stated that the ascent of sap is modified in the same way by different conditions such as anaesthetics.

Such being the case, he had no diiliculty in formulating his theory that the ascent of sap is effected by the rhythmic activity of living cells. It should be noted," as Dixon has pointed out in his admirable criticism of this work *The Transpiration Stream,* "the peculiar verbal transitions which often form an essential part of his arguments. Here alternate contractions, varying from one to four minutes or so, become ' pulsations,' and these glide easily into ' rhythmic activity.'

An experiment by Bose in support of his views is quoted here. He employed an electrometric method depending on his idea that " the electric condition of the tissue undergoes a definite variation under changes of turgor ; a diminution of turgor introduces an electric change to galvanometric negativity, while an increase in turgor, on the other hand, an electric change to galvanometric positivity."

Now a plant under drought has its turgor diminished throughout its length. If then, two electrical contacts are made, A with the stem and B with a distant leaf, the electric condition of the two points will be more or less similar. Addition of water to the soil, however, will cause an ascent of sap, which will reach the lower contact first and will induce an increase of turgor at this point.

This will at once be signalled by a sudden change to positivity at A. This experiment was repeated by Dixon, aided by Dr. J. H. Poole, using a galvanometer which gave a deflection of 1 mm. for 0-004 volt. The plants employed were *Sparmannia africana* and *Pelargonium zonale.* Although the soil round each was dry, neither showed any signs of wilting. " After connection had been made as described by Bose, a slow movement of the spot of light was recorded. A few minutes later the spot came to rest at 19-7 mm. to the left of zero.

This, as the instrument was arranged, indicated that the point A was positive relative to B. When observation showed that the spot was sensibly stationary, the earth in which the plant *(S. africana)* was growing was irrigated with water at about 25° C. Immediately the spot moved to 17-0 mm. and slowly crept back to 19-3 mm., which it reached two minutes later.

During the next thirty minutes it crept slowly to 21-0 mm. During all this time nothing like the to and fro motion, corresponding to Bose's pulsations, was observed, nor was there any sudden deflection. Point A was 3 cm. above the surface of the soil.

The temperature of the room was 14-0° C." Similar results were obtained for *Pelargonium zonale.* Dixon also failed to confirm Bose's statement of a rhythmic change of potential in the cells of the stem by means of Bose's method of the electric probe. This " electric probe " consisted of a platinum wire coated with celluloid and enclosed in a glass tube. This glass tube was fixed firmly in the body of a microscope, so that the sharp point of the wire, from which the celluloid had been stripped, projected from the glass tube and occupied the position of the objective.

With this instrument Dixon attempted to repeat Bose's results. The body of the microscope was placed horizontally and the vertical stem of the experimental plant was firmly clamped to the stage. In this way the point of the probe was gradually advanced through the tissues of tin stern by the action of the micrometer screw of the microscope. By means of mercury in the heathing glass tube, connection was made between the probe and one pole of a galvano-meter, while the other pole was connected with another platinum contact embedded in the tissue of a morbid leaf.

The circuit was completed through the plant. With this apparatus and using a highly sensitive galvanometer giving a deflection of 1 mm. for a current of $10''^{10}$ amperes, Bose recorded that the galvanometer showed alternate deflections in opposite directions when the probe had advanced a certain distance inwards. The period of this alternation varied according to circumstances from 13-5 seconds to about three minutes. The amplitude of the vibrations varied between 10 mm. and 66 mm. Dixon, with a galvanometer giving a deflection of 1 mm. for a current of 5 X 10^{-10} amperes, and using *Pelargonium zonale, Impatiens sultani* and *Chrysanthemum sinense,* found that there was a slight " kick " every time the probe was advanced 0-1 mm. into the tissue at intervals of ten minutes, but in no case was there any regular vibration recorded.

Bose considered that the osmotic activities of the leaf cells are too slow to account for the rapid rise of water in plant stems. " He quite fails to realise that slow osmotic activities spread over the large area of the bundle terminations would cause a comparatively rapid rise of water " (Dixon). Bose's criticisms of Strasburger's work are entirely beside the point. He stated that the living cells above the dead portions of the stem are able to pump water through the vessels, and thus " unconsciously admits the substance of the cohesion theory ".

Benedict has determined the applicability of Hose's theory by comparing the maximum rate of flow which is possible according to his theory with the known rate of flow. The cross-sectional area of the living cells in the trunks of ten trees of different species was measured, and the maximum pumping capacity at the most rapid pulsation rate found by Bose, was computed. The actual rate of sap flow under maximum conditions of transpiration proved to be from eight to thirty thousand times as rapid as would be possible under Bose's theory, *Quercus rubra* giving the first value and *Robinia pseudacacia* the second, while the other eight species used presented an intermediate condition.

Chapter 8

Translocation

In the higher plants the food material elaborated during active metabolism must be transferred from one part of the plant to another for purposes of storage. Thus, sugars synthesised in the leaves are removed to other parts of the plant body and stored as insoluble polysaccharide for future use. Similarly, proteins have also to be translocated away. Again salts from the soil solution must be passed up through the plant for a variety of anabolic purposes. It follows, therefore, that paths of passage must exist in the somatic organisation of the plant, so that this process of translocation may be facilitated.

It is important to bear in mind that translocation of material is regulated by supply. Thus substances present in sufficient quantity may undergo very rapid translocation until the needs of the plant are satisfied. Moreover, different substances may be translocated in different directions at one and the same time.

In the germinating seed, the reserve stores of food in cotyledon or endosperm have to be conveyed in two different directions:

1. To the upwardly developing shoot.
2. To the downwardly developing root.

In the case of annual plants growth ceases towards the end of the season and translocation of elaborated food material becomes a serious problem, inasmuch as the ripening fruit and seeds must be supplied with sufficient reserve stores for successful germination in the following season. In annual plants, then, there is an upward translocation to maturing fruit and seeds, while in perennials the reverse holds good, for the elaborated food is passed down to the 11ndcrground rhizomcs, roots, bulbs and similar organs for storage during the winter months of the year.

They arc utilised from this source when growth recommences in the spring. Translocation is now in the reverse direction, and the stored food, after it is brought into a suitable state for removal by the action of hydrolysing enzymes, is carried upwards into the newly developing buds.

In the higher plants the internal tissues may be roughly divided into wood or xylem which is dead tissue, and the bast or phloem as well as ground parenchyma, which are living cells with protoplasmic contents. The question arises which of these tissues carries away the elaborated food material from the active centres of synthesis.

The older investigators considered that the function of the xylem was to carry water and dissolved salts, while the phloem furnished the channel for the downward passage of the elaborated products of metabolism. The chief evidence for the phloem being the channel of transport of the already synthesised organic food material was largely based on ringing experiments. It has long been known that if a stem were " ringed," then the passage of food apparently ceased.

The main line of translocation was considered to be through the sieve tubes ; the companion-cells and phloem parenchyma merely played a subsidiary part. Thus the products of photosynthesis and protein synthesis were primarily mobilised in the leaf and were then removed to the storage organs *via* the sieve tubes.

The early work of Czapek appeared to give complete confirmation of this picture translocation and his experiments are too well known to need more than the very briefest reference here. Czapek, showed that if the petioles of leaves were killed, either with steam or with chloroform, the depletion of carbohydrates was completely stayed, or if the petioles were cut on one side there was considerable delay in the removal of carbohydrates in that half of the leaf. On the other hand, if the petioles were immersed in a 5 per cent, solution of potassium nitrate, carbohydrate transport was not inter-fered with.

He therefore concluded that plasmolysis of the phloem did not interfere with the transport of sugars. He considered the sieve tubes to be the chief organs of transport because the deposition of callus synchronised with the stoppage of translocation.

It was not until 1911 that any serious criticism was brought against this work. In that year Deleano stated that if the petioles of leaves were killed, even in such circumstances, carbohydrate still left the leaf, although there was considerable reduction in the rate.

Dixon, in the course of his address, pointed out the well-known fact that Fischer had found reducing sugars as well as proteins to be present in the xylem, and further that he (Dixon) and Atkins had shown that the sugars in the wood were not only hexoses, but that disaccharides such as sucrose and maltose were also present, and that these carbohydrates were to be found in the tracheae at all times of the year.

Dixon considered that ringing experiments are of little value, inasmuch as the surface of the wood may be injured and more important still, may become blocked with air-bubbles into which substances are exuded by morbid changes in the cambium, medullary rays and wood parenchyma.

However, the more cogent of Dixon's arguments in this connection lie in certain calculations that he has made with regard to the actual quantities of organic material involved in transport and the velocity of flow in the channels which are necessary to effect this transport. He gives as an example a selected potato tuber of weight 210 gm. The total cross-sect ion of the bast in the slender branch was 0.0042 $cm.^2$ no allowance having been made for the thickness of the cell walls. In 100 days all the material in the tuber must have passed through this cross-section.

On analysis it was found that 24 per cent, of the tuber was combustible, therefore more than 50 per cent, of carbohydrate passed through this conduit. As a matter of fact, the concentration is probably never higher than 5 per cent, and possibly not more than 4 per cent. Dixon considered that the nature of the sieve tubes with their thick, viscous contents, do not admit of such a rapid rate of flow. He further thought that there is a downward as well as an upward flow of water in the xylem and cited two experiments in support of this view. A petiole of *Sambucus niger* was split longitudinally and one half removed and the attached portion was then placed in a solution of eosin.

It was found that the solution was rapidly drawn up the wood capillaries of the intact half petiole and soon appeared in the veins of the pinnae on the same side of the leaf, beginning with the lowest and gradually working up into the upper ones, and finally into the terminal pinna. In the next stage the dye worked its way through the veins of the pinnae on the other side, and eventually to the lowest pinnae on the split side of the petiole. The second experiment of Dixon consisted in cutting the tip of an upper leaf of a potato plant under a solution of eosin. The liquid was quickly drawn back into the tracheae of the leaf and passed down the stem even into the tuber.

Both experiments, however, are open to criticism. In the case of the leaf of *Sambucus niger* it would be expected under the experimental conditions employed that the dye would be drawn up one series of pinna' and would eventually pass down the other side. This, therefore, is no proof of the reversal of the transpiration stream. That a backward suck was exerted by the potato leaf is also not surprising if Dixon's cohesion theory is correct. One of the main points of the cohesion theory is the fact that the water is considered to be under a strong tensile strain. In such circumstances, if the strain at one point were broken in any way, the strain under which the water is suffering in the other tracheae would tend to cause a suck back and show apparent reversal where in reality none existed.

According to Dixon, translocation can be best explained on the cohesion theory of the ascent of sap. " Transpiration, by drawing off water from the tracheae, causes water filling these tracheae to pass into a state of tension. This tension determines a flow from any source wherever situated and the continuous column in one series of tracheae to draw down solution in a neighbouring filament of tracheae terminating above in some local supply."

Although Dixon's view that the xylem and not the phloem was the channel of transport was accepted for some time after its announcement, there nevertheless has been a growing opposition to the theory, and the most recent work on the subject shows quite definitely that it is the phloem which is the channel of transport of carbohydrates. Arndt has reported that there is an upward asovell as a downward movement of eosin solution in the same ring of xylem in the coffee tree.

The solution usually moves in the outer ring of the xylem. One-sided staining, discovered when the eosin was applied to the lateral roots, indicates that specific roots are connected with particular branches of the crown. In daylight there is a more rapid upward current of eosin, and it is only in the late evening, at night and the early morning, that the downward movement is not greatly reduced. Arndt considered that this downward movement is quite adequate to account for the downward transfer of solutes.

Curtis found that in ringed stems protected with paraffin wax there was a considerable interruption of the upward supply of carbohydrates as well as nitrogen and ash to the growing regions of plants, whereas only very small amounts of xylem were necessary for the supply of water.

By the use of divided stems whereby water was supplied to the top by one set of roots and nitrogen by another, Curtis was able to show that the nitrogen was not transferred by the xylem, while if the xylem were connected by a short strip of phloem then a normal supply would be translocated upwards. The experiment, however, was not of a very satisfactory nature, and by a later series of experiments, he was able to show that the xylem alone was not satisfactory as a channel of transport for nitrogen, while the phloem alone was quite as satisfactory as xylem and phloem together.

Dixori's calculations based on the potato tuber is certainly a difficult point to explain, but as Curtis showed, no sugars have ever been found in the xylem of the potato, and further, measurements based on the work of Artschwcger indicate that the cross-section of the xylem is smaller than the cross-section of the phloem of the tuber.

Curtis agreed with the view put forward by de Vries many years ago that the rotation of the protoplasm helps in the conduction of food down the phloem, and he considered that protoplasmic rotation is not a pathological phenomenon as some investigators have assumed, and cited the cases of *Nitella* and *Elodea,* where the streaming exhibited by the protoplasm is. undoubtedly a normal phenomenon of the cell.

If, according to Curtis, the phloem supplies an efficient mechanism for the upward supply of carbohydrate, nitrogen and ash, then it is probable that it is just as efficient for the downward transport of metabolic products from the leaf.

Weevers, in order to determine whether the upward transfer of foods during periods of terminal shoot growth takes place through the xylem or the phloem tissues employed ringing experiments on the variegated branches *ofsEsculus Hippocastanum* and *Acer Negundo.* The use of branches bearing leaves with no chlorophyll eliminated the possibility of the manufacture of food above a ring, which would occur in normal green leaves, thus obviating the necessity for defoliation or for darkening of this part.

Ring wounds were protected with wet bandages or the application of cocoa butter. Experiments were carried out in the spring just before the buds started growth and also in midsummer. The experiments consisted of the complete ringing of green and yellow shoots and the partial ringing of yellow shoots. In either spring or midsummer, complete ringing of the shoots prevented normal growth, while partial ringing of the yellow shoots or complete ringing of normal green shoots allowed considerable growth to take place.

Analysis of ^*Esculus Hippocastanum* showed 3 per cent, of reducing sugars in the green leaves, 1 per cent, in the yellow from unringed branches and only traces from the ringed branches. When ringed, the leaves from the yellow branches often showed early withering and death. Though no variegated branches were available, comparable experiments were conducted with *Nerium Oleander,* which bears phloem strands internal to the secondary xylem.

Ringed branches, darkened by enclosure with blackened paper, continued growth indicating that ringing in this case had not prevented food movement. This work shows very clearly that removal of the phloem, contrary to Dixon's statements, does very seriously interfere with the translocation of carbohydrate.

The most conclusive work on this subject has recently been carried out by Mason and Maskell in Trinidad on the cotton plant. This investigation based entirely on quantitative data can only be briefly dealt with here. For further details the original two lengthy memoirs should be consulted, which are admirable examples of lucidity of expression combined with care in the interpretation of results.

Mason and Maskeli estimated the quantity of carbohydrate in the bark, leaves and wood of the cotton plant at different times of the day. They expressed their results as " residual dry-weight," *i.e.,* total dry-weight less carbohydrates. It was discovered that the greatest changes in amount of carbohydrates occurred in the bark and leaves, whereas in the xylem it remained more or less constant in amount. Stems were also ringed and the exposed portions protected with vaseline to prevent external changes.

It was found that there was an interruption of carbohydrate supply across the gap. The ringing, however, did not interfere with the passage of carbohydrate into the stem immediately above or immediately below the ring. An accumulation of sugars was found in both bark and wood just above the ring and also in the leaves 2 feet away.

Yet throughout all these experiments the concentration of sugars in wood remained fairly constant. Further experiments with ringed stems in which the flaps of bark were separated from the wood by means of paraffined paper, showed that conduction took place at practically the normal rate. Hence connection between bark and wood is not necessary for normal conduction.

Diurnal estimations of the sucrose supply to the bolls showed that translocation was four times as fast by day as by night. The variations of sugar transport into the bolls could be significantly correlated with variations in the sucrose gradient from bark to boll, and it is perhaps by means of these gradients of concentration that the supply of sugars is removed from one part of the plant to another.

Samples of bark taken from different levels of the same stems showed that there was a concentration gradient in the bark down the stem towards the roots which resembled the movement of diffusion in that the direction of movement was from a region of high concentration to one of low concentration. The major part of the total fluctuations of sugars in the bark is due to sucrose, and it is therefore suggested that the bulk of the carbohydrate travels in this form.

The main significant difference between physical diffusion of sugar in water and the movement of sugar in the plant is that this movement is enormously greater in the latter. It is, in fact, forty thousand times as great as the diffusion constant in a 2 per cent, solution of sucrose in water, and is almost identical with the diffusion constant for molecules the size of sugar molecules diffusing in air.

Although Mason and Maskell use the term " bark " for the channel of transport, it must be understood that this term covers all the tissues outside the xylem. The main flow of the carbohydrate was discovered to be in the sieve tubes. The question therefore arises as to whether any mechanism in the sieve tubes can be visualised which would reduce the resistance by the medium to diffusion.

Tempting in the circumstances to postulate the existence in the tubes of some special organisation of capillary structure which greatly reduces any resistance to diffusion. Nevertheless, the simple fact remains that we arc quite ignorant of any such structure. A probable mechanism, at least from the physical standpoint, consists in the circulation of the contents of the sieve tubes. A chain of sieve tubes would then form a scries of vortices and exchange of solute between vortices would follow the concentration gradient between the vortices.

All the characteristics of the phenomena of diffusion, except the absolute magnitude of the rates, would be shown by this system, so that it would satisfy the known facts with regard to the movement of sugars in the plant. It is evident that there are two aspects of this mechanism. There is first the movement within the sieve tubes, and secondly the movement from sieve tube to sieve tube.

It is suggested by Mason and Maskell that the former might be brought about by the streaming of the contents of the sieve tubes. That is to say, that the concentration within the sieve tubes is rapidly equalised by. a mass movement of the contents. The difficulty is that such a movement of the contents of the sieve tubes lias not yet been observed, for the streaming of the protoplasm apparently ceases in the mature tubes. It seems also that the rate of this movement, if it should occur, would have to be greatly in excess of the rate of protoplasmic streaming observed in other vegetable cells.

Curtis has found that when the petioles of *Phaseolus vulgaris* are chilled between 1° C. and 4° to 6° C., the removal of carbohydrate from the leaf blades is greatly retarded or even completely stops. ~ The temperatures which cause a check in trans-location are approximately the same as those causing a cessation of protoplasmic streaming, a fact giving additional support to the theory that protoplasmic streaming is intimately connected with solute passage in the phloem.

Chilling also interferes with the upward passage of inorganic material, and if the petioles are enclosed in tubes containing nitrogen under slight pressure, translocation ceases. On the other hand, removal of carbohydrate from the leaves does not appear to be appreciably interfered with when the petioles are coated with wax. This may be due to the presence of a petiolar cavity which still allows of effective aeration. This w^rk indicates that upward as well as downward transloca-tion of solutes, organic and inorganic, is brought about by the activity of living cells.

Whatever the actual mechanism may be, the fact remains that Mason and Maskell have demonstrated bcvond all doubt that the transport of carbohydrate is *via* the phloem, and not as Dixon considered through the xylem.

Although the path of carbohydrate transport in the plant has been so successfully traced by Mason and Maskell, the problem of nitrogen transport is still in an uncertain state. The usual view on the subject is that nitrate from the soil is absorbed by the roots and is moved upwards in the transpiration stream through the xylem to the leaves, where it is synthesised to amino-acids and proteins. The elaborated products are exported down the «tem *vid* the phloem. The whole problem, however, is complicated by the investigations of Curtis, who claimed that notonly is organic nitrogen transported by the phloem, but also mineral nitrogen.

According to Chibnall, there is a diurnal variation in the nitrogen-content of the cabbage leaf, the total nitrogen increasing by day and diminishing by night. It may therefore be concluded from this work that there is a synthesis of nitrogen products in the leaf by day and export of nitrogen by night.

Maskell and Mason reinvestigated the whole problem of nitrogen transport in the cotton plant. The results were expressed both as fresh weight and also as residual dry weight. The nitrogen was estimated in the leaves, bark and wood at different times in the day and night. Expressed on a fresh weight basis, the diurnal variation in nitrogen-content of the leaf followed a definite pattern. Statistically considered, the variation in nitrogen-content of the leaf showed a definite variation with time, whereas, in the case of the bark, the observed standard deviation due to sampling was not much greater than the standard deviation in time.

Therefore, statistically, there was no real evidence of variation in time in the nitrogen-content of the bark tissues. Inspection of the curve suggests that the changes in the bark tend to lag behind those in the leaf. It is possible that concentration changes in the leaf lead to similar changes in the concentration of diffusible nitrogenous substances in the bark sap, and that consequently physical diffusion may be a factor in the transport of organic nitrogen from the leaf. I

Expressed as residual dry weight, the results are not so marked as in the case of the fresh weight values. Nevertheless, the night values are less than the day values by both methods, and indicate withdrawal of nitrogen by night. When free boiling is taking place the nitrogen-content of the bark just below the foliage region shows no appreciable diurnal variation. This is possibly due to the fact that the bulk of the nitrogen passes into the bolls and little to the roots.

Ringing the stems causes an accumulation of nitrogen in both bark and wood above a ring, and decrease in both below a ring. There also appears td be a horizontal movement of nitrogen from bark to wood. Estimation of nitrate nitrogen shows that ringing does not prevent its entry into the leaf. Moreover, the downward movement of nitrogen is not interfered with by the separation of the bark from the wood.

A consideration of the data as a whole presented by these authors appears to be in accord with the original view of nitrogen transport, that is to say that the bulk of the mineral nitrogen absorbed by the roots is transported up the stem in the transpiration current through the wood, while the elaborated nitrogen is removed from the centres of synthesis *via* the phloem.

THE PASSAGE OF STIMULI IN PLANTS

The reaction of plants to stimuli appears to be connected with certain specific substances of the nature of hormones. Hormones were first postulated by Bayliss and Starling a number of years ago to account for definite animal reactions and were brilliantly verified since, by the discovery and isolation of some of these bodies, such as thyroxin, insulin and adrenalin. They appear to play a *role* in the physiological activities of plants.

The best known case of hormone activity and the one that has been most fully investigated, is that of the sensitive plant, *Mimosa pudica*. The leaves are very sensitive, and close on touching in a very characteristic manner. It was first shown by Ricca that the stimulus or hormone in *M. pudica* passes through the xylem and not through the phloem according to Haberlandt. In the animal economy there are two methods whereby a stimulus may be perceived:

1. Through the nerves, which form a rapid method of signalling and can be likened to a telegraphic system
2. The stimulus may be transferred by a hormone.

The latter is a slower method compared with the former, and may be likened to a postal system. In plants, no structures comparable to nerves have been discovered, and although it is equally true that no hormone has as yet been isolated from a plant, nevertheless, from their responses, the presence of hormones must be inferred.

In *Mimosa,* it was shown by Ricca that if the stem were killed with steam or separated by a water-gap, the hormone still passed in the upward or downward direction. This work was further extended by Snow, who confirmed Ricca's original observations and added several of his own. He showed that though the phloem of the stem was quite insensitive to the passage of the hormone, the hormone was conducted through the phloem in the petiole, the so-called " high-speed conduction."

Although the hormone is thermostable in the plant, watery extracts are killed by boiling. Snow prepared water extracts of the hormone, and found that it was not precipitated by lead acetate and that it gave 110 protein reactions. It was able to diffuse through a collodion thimble without losing its properties. It is possible that it may have a comparatively simple chemical constitution. Could it be obtained in sufficient quantity a study of its chemical properties would be of the utmost interest.

Ball has also investigated the conduction of the stimulus in *Mimosa,* and has demonstrated that at least two methods are in existence. The normal method of conduction through the xylem, such as was described by Ricca and Snow, in which the stimulus travels at the rate of about 15 to 28 cm. per minute, was confirmed, and a second still more rapid method of conduction was found to exist in the pith, where the rate of passage is 200 cm. per minute.

This latter rapid conduction method is shown when cut shoots are completely submerged in water. Ball proved quite clearly that the stimulus did not pass through the phloem, for on decortication and removal of all tissues external to the wood, the stimulus still passed with the same rapidity as before. Under normal conditions the hormone is released at the point at which the stimulus is applied and is transported by the water current through the xylem, and this method of conduction operates best when the turgor of the cells is low and the water tension in the vessels is high.

When the turgor of the tissues is high, the hormone is still released at the point of stimulation, but, instead of passing up the xylem, it merely causes contraction of the neighbouring cells which are possibly situated in the pith. These in turn liberate a further quantity of hormone, and so a relay mechanism of a highly efficient nature is set up, whereby the stimulus can pass in either direction in the plant and is independent of the water current.

Ball has also ascertained that both methods of conduction can operate at the same time. The latter method of conduction (*i.e.*, through the pith) is probably of the same nature as the " highspeed conduction " described by Snow, due to the almost explosive contraction of highly turgid cells ; the chief difference lies in the fact that high-speed conduction is through the phloem of the petiole.

The passage of phototropic and gcotropic stimuli has recently been shown to be probably due to the transference of hormones. The coleoptile of grasses has been considerably investigated with respect to its response to light, and a very large body of data has now been accumulated on the subject.

De Candolle put forward the suggestion that the bending of an etiolated shoot towards incident light was caused by the fact that the highly illuminated side grew less than the side away from the light and a curvature towards the light was thus developed. In other words, the light had a direct effect on the plant, and was directly responsible for the resulting curvature.

On the other hand, Pfeffer considered that light acted as a releasing mechanism and there was no quantitative relationship between the externally supplied energy and the resultant response of the plant. The stimulus, in this instance light, acted like the trigger of a gun. There is no quantitative connection between the pulling of the trigger and the ensuing explosion. It does not matter whether the trigger be pulled quickly or slowly, hardly or softly, the force of the explosion is always the same, and depends on the amount of powder in the cartridge.

This conception of a stimulus affords the best explanation of the results of Charles Darwin with the etiolated coleoptile of various grasses. Darwin discovered that the response of an etiolated coleoptile depends on the perception of the stimulus (in this case light) by a localised region of the organ *which is not the region of curvature.* This work has been confirmed again and again by numerous investigators.

The curvature is always present at the base, if the tip of the coleoptile be illuminated. Hence there must be a perceptive and a responsive region. As a matter of fact, it has been discovered that only about 2 mm. of the apical region is sensitive. It is therefore evident that the excitation is released at the tip and travels down to the motor region at the base where it produces curvature.)

Morphologically, the coleoptile is a cylindrical structure with two lateral veins. Rothert cut the veins and discovered that the stimulus still passed when the tip was illuminated. This observation was confirmed by Fitting. Boysen-Jensen, however, obtained a different result. He ascertained that if in the air of the laboratory, the vein away from the light were nicked, there was no response. Similarly, if the vein towards the light were cut there was a positive response, *i.e.*, the coleoptile bent towards the light.

He confirmed the result of Rothert and Fitting that if the coleoptile were kept in a saturated atmosphere and the vein away from the light were cut, there was a positive response, and the same result was obtained if the vein were cut under water. Boysen-Jensen therefore concluded that the saturated atmosphere kept the gap full of sap and the substance or substances responsible for the response diffusing down from the apex could pass through the water-filled gap. If a mica plate were inserted in the cut made away from the light, there was no response, while if the mica plate were inserted on the side towards the light, curvature still occurred.

These results have been amply confirmed by Purdie and Nielson. Bovsen-Jensen next cut off a centimetre of the tip of the coleoptile and replaced it by gelatin base. He discovered that on illumination the coleoptile still responded. It is thus evident that the substance responsible for curvature can diffuse through gelatin. This experiment has also been independently confirmed.

Paal has made a number of investigations of coleoptile curvature and considerably modified Boysen-Jensen's original technique. He has shown that none of the results obtained arc due to stray light from the illuminated apex. A further experiment of Paul's is important in this connection. He removed the tip of the etiolated coleoptile of *Coix* and replaced it without gelatin in the dark. A curvature occurred through 90° with the side of the stump beneath the tip on the outer convex side of the curve. If, however, the apex were replaced with gelatin, no curvature occurred when the plant was replaced in the dark.

Paal concluded from this experiment that there must be two substances or two types of substances diffusing from the tip. One of these promoted growth, but was incapable of diffusing through gelatin, while the second, produced by the stimulus of light, was capable of diffusing through gelatin and was responsible for the curvature.

Boysen-Jenscn considered these substances diffusing from the apex to be growth promoting, and from his experimental results supposed them to pass down in greater quantity on the side *away* from the light. Paal, on the other hand, held that they are perhaps destroyed on the side towards the incident light as a result of illumination.

A series of experiments by Stark has shown that the apices of foreign genera placed on stumps also give a response. If the tip of an *Avena* coleoptile were placed on the stump of *Hordeum,* a response was still given, although the reaction was much weaker in the majority of cases. Yet in the particular case quoted here, *Arena* tip or *Hordeum* stump, the response was frequently much stronger than if the decapitated tip of a *Hordeum* coleoptile had been replaced on its own base.

Stark also discovered that it did not matter whether the apex were accurately replaced or even if it were replaced back to front; the response was always similar. Stark further tried the effect of placing extracts from wounded coleoptiles on the side of the stump and found that response occurred towards that side.

Stark and Dreschel have also tried the effect of decapitating coleoptiles and illuminating the tips with unilateral illumination. The tips were then replaced on the stumps, and it was found that a response was given in the direction of the original illumination of the tips. An important experiment relating to this work of Stark and Dreschel has been carried out by Brauner, who removed the tips of the coleoptiles and placed them in the dark and exposed the stumps to unilateral light.

The tips were then carefully replaced on the stumps and the plants replaced in the dark for four hours. It was found that twenty-one out of the thirty-two coleoptiles used showed a curvature in the direction of the original light. The control plants, *i.e.* those in which the tips were not replaced on the illuminated stumps, showed no response at all.

Priestley objected to the above very positive evidence to prove that the action of the light is to release some substance or substances from the apex, which travel to the base and there bring about a curvature. He considered that since the coleoptile possesses limited growth the curvature exhibited can be explained on simpler grounds. During the period it is employed in photo-tropic experiments its increase in length is entirely by cell wall extension, no new cells being formed, and since cell wall extension is more rapid in the light than in the dark, though the total increase in length is less, positive curvature occurs.

The fact that Blaauw, with a spor-angiophore of *Phycomyces nitens,* equally illuminated by an arrangement of mirrors, the lowest intensity of illumination of which was 210 metre-candle-seconds, showed that no increase in growth occurred for two to four minutes and that this latent period was followed by a marked acceleration in the growth rate. The greater the intensity of the light the less the time of the initial latent period.

Yet a sporangiophore submitted to unilateral illumination shows a positive curvature, whereas from Blaauw's results a negative response should have been shown. Blaauw has given the explanation that this was due to a " lens-action " on the part of the sporangiophore.

This structure is cylindrical in cross-section and transparent and Blaauw held that the light was brought to a focus on the side of the sporangiophore away from the light with the result that an acceleration of growth occurred on that side, showing a positive curvature. Priestley therefore considered that there is a greater quantitative relationship between incident light and response which negatives the suggestion of stimulus in one region and response in another.

Apical guttation also presents difficulties to Priestley, who stated that the localisation of an apical hydathode explains the direction of the usual autonomous curvature of the coleoptile and he would suggest that the permeability of the coleoptile tissues are increased by light and therefore light falling on the apex will increase apical guttation. Any lateral light that may fall on the apex, therefore, increases the rate of flow from the vein nearer the light and produces phototropic curvature. When the apex is removed guttation is very free and growth much reduced until the cut surface is blocked.

If the apex be replaced, the guttation is immediately reduced and growth is greater. If the apex be replaced on the side of the stump only, this side alone is blocked and curvature results. Should the replaced apex have been laterally illuminated, the block is less complete on this side of the stump and positive curvature results.

This theory is too full of a mass of suppositions and assumptions to have any value attached to it. Why precisely a greater flow of sap on the side towards the light (and this is a pure assumption with no experimental foundation) should cause a positive curvature is difficult to say. Priestley's views are still further discredited by the recent important paper by Went, establishing the existence of some growth substance in the coleoptile of the oat.

A number of coleoptiles were grown in darkness and all subsequent manipulations were made upon them in red light which had no effect. The tips were removed and placed upon thin slices of jelly (either gelatin, agar or silicic acid). If a portion of this jelly were replaced asymmetrically on the stump of another coleoptile from which the tip had just been removed, a negative eurvature was found to occur ; in other words the coleoptile bent away from the side receiving the jelly. Went interpreted this result as being due to the hormone diffusing from the jelly into the base of the coleoptile. At 25° C. in 170 minutes' time the negative curvature is followed by a positive curvature.

This is considered to be due to the regeneration of a new physiological tip, which it would seem is first formed on the side of the decapitated stump away from the jelly, and from here a new supply of growth substances is released into the base of the coleoptile. Went has further shown that the amount of curvature is proportional to the number of tips that were originally placed upon the slice of jelly and to the period of time that they are allowed to remain upon it.

Went suggested that the action of the hormone is to modify the process of extension in cell size. On the other hand, the substance appears to be incapable of extending the size of the cells at the apex of the coleoptile. He therefore suggested that there is a second factor that must be taken into account, namely, materials capable of causing cell wall extension, which must be produced in the base of the coleoptile.

Went has made an attempt to identify the mode of transport of the hormone. He found protoplasmic streaming occurring in the cells of the coleoptile and considered this phenomenon to be responsible for the downward transport of the hormone. He has even watched the transference of the substance from a piece of jelly at the top end of a coleoptile stump down the coleoptile to another piece of jelly at the base. Moreover, from rough diffusion experiments he has been able to arrive at the approximate molecular weight of the hormone and found it to be about 376.

For further details of this interesting work the original memoir should be consulted. Went's experimental technique is above reproach and leaves no doubt that the response in the coleoptile is due to a hormone and that Priestley's somewhat crude mechanical explanation has no foundation in fact.

It is well known that if the apices of many seedlings of mono-podial growth are removed the axillary buds grow out and form new shoots. The apex exerts some influence which prevents the development of the axillary shoots. The suggestion has been made that an inhibitory substance is formed at the stem apex and is passed down the shoots which prevents the development of axillary buds.

Another suggestion is that the growing apex in some way draws to itself the supply of nutrient fluid or perhaps some particular substance necessary for growth, and the axillary buds are thus kept in the dormant condition through starvation. Loeb in a long series of publications on the regeneration of *Bryo-phyllum calycinum* considered inhibition to be due to the lack of necessary nutrient material.

Snow has critically examined Loeb's work, and considered that the transport of an inhibitory substance or hormone from the stem apex to be the true explanation. Using *Phaseolus multiflorus* and *Vicia faba* he was able to show that ringing the epicotyl, so that all tissue including the cambium was removed external to the wood, still allowed of the passage of the inhibition. The control plants which were decapitated and unringcd showed signs of regeneration after four days, whereas the treated plants showed no axillary growth at all.

It is thus evident that the inhibition can pass through a stem that has been ringed. He also discovered that the inhibition could pass through the pith, although it is considerably weakened in its effects. When the main apex and axillary bud were only connected by the xylem with a few cells of pith parenchyma adhering to it, inhibition again passed through. In a second series of experiments a zone of the stem of *Vicia faba* was killed by a jet of steam playing on it for twenty to thirty seconds. Although only the outer cortical cells were killed by this treatment the inhibition failed to pass owing to " physiological shock " and the axillary buds grew out below the treated zone, the main stem continuing its growth.

When the cut surfaces of tissues of seedlings were bound together in pairs and each pair was decapitated the inhibition was not able to pass across from one to the other. The question as to whether the inhibition could or could not pass across a watery gap proved difficult to show experimentally. The difficulty, however, was surmounted by splitting *Phaseolus* seedlings longitudinally through the lower region of the epicotyl, cotyledons, hypocotyl and main root and binding the parts together. The inhibition was found to pass through the watery gap.

To test whether the inhibition passed upwards in the transpiration stream, young seedlings of *Vicia faba* were decapitated in the epicotyl, so that the two axillaries of the cotyledons grew out. Those seedlings were then selected in which the axillaries were nearly of equal strength and the shorter of the two was then decapitated above its second leaf and a bud was allowed to remain in the axil of one of the two leaves of this shoot, while the bud in the axil of the other was removed.

The remaining bud on the shoot showed scarcely any growth even after three or four weeks. Presumably therefore it was inhibited by the apex of the longer shoot, for if this were also decapitated then the bud grew out strongly. A one of tissue (about 4 mm. long) was now killed by a hot glass rod without charing the tissues or causing the cell-contents to boil and the dead zone was painted over with vaseline. The controls, in which the apex of the longer shoot was also removed as well as the axillary buds, save the one to be measured, showed a mean growth of 3-37 mm. compared with 1 -54 mm. for the treated plants. The difference of the mean is 1-83, which is 4-30 times its own standard deviation and therefore fully significant.

The interpretation placed by Snow on these experimental results is that an inhibiting substance formed at the growing apex of the longer shoot is transported through the living tissues and leaks into the xylem just below the dead zone. It is then drawn up with the transpiration stream through the dead zone and leaks out into the living tissue of the shorter shoot where it inhibits the growth of the axillary bud.

ROOT-PRESSURE

The excretion of fluid is a common phenomenon throughout the plant world and may take place either from the intact plant or after injury, as in the bleeding of cut stems. This exudation of water falls into two well-marked types, which are termed respectively *bleeding* and *guttation.* In the case of bleeding, it is well known that if the stems of many plants are cut, especially in the spring, there is a copious flow of liquid which is excreted under considerable pressure. This is usually spoken of as *root-pressure,* but a better term would be *bleeding-pressure,* for although it is undoubtedly true that roots excite the most active exudation of water, this is merely because they are the best absorptive organs the plant possesses.

Stock does not, as a matter of fact, always bleed more actively than portions of the stem and in a number of cases it has been found that active bleeding will take place in stems severed from their root-stock. The term root-pressure " is therefore something of a misnomer in this respect and is better replaced by the term bleeding-pressure." The term ' guttation covers the exudation of drops of fluid from the uninjured parts of plants, such as the leaves, hydat nodes, nectaries and similar structures.

The fluid that escapes in bleeding may be either practically pure water or may contain considerable quantities of organic matter. The sap is exuded from the xylem, either tracheids or vessels. Bleeding, therefore, corresponds with filtration under pressure and the exuding liquid may or may not be mixed with bubbles of air.

Bleeding is a vital phenomenon and is markedly affected by external conditions. A dry soil or active transpiration may cause considerable diminution of bleeding, and anaesthetics such as chloroform also exert a considerable effect. When the temperature is too low bleeding ceases, although the absorption of water still continues. Finally oxygen is a necessary factor, and in its absence bleeding is brought to a standstill.

Chapter 9

Nutrition of Plants

The intake of manurial or nutrient salts is carried out in the plant in the same region of the root as water intake, namely, the region bearing root-hairs. Before intake is possible the salts must be in solution, and therefore free to diffuse. If the molecules of any one solute are to diffuse into the plant there must be more of that kind free outside the cell than there are inside.

A concentration of any particular kind of molecule greater outside the cell than inside does not invalidate what has been said in regard to water intake. Just as the water molecule concentration in a solution is lowered by all the solute molecules present, so is the concentration of any one solute reduced by all the other solutes *and* the water.

The key to the problem lies in the statement that each kind of particle diffuses on its own gradient. The direction in which water molecules are diffusing does not affect the direction in which a solute molecule is diffusing, except in so far as either affects relative dilution.

SOLUTES INTAKE

The protoplasmic lining of the cell is an imperfect semi-permeable membrane and is permeable to many solutes. Hence, if the protoplast is permeable to a given salt molecule (or ion), and there is a gradient of concentration inwards of that salt, entry is effected. These two factors—degree of permeability and the presence of a gradient—are of fundamental importance. The percentage concentrations shown refer to numbers of individual molecules. The arrows indicate the direction of diffusion in each case. At one time or another nearly all of the known elements have been found during the analysis of plant material.

This does not mean that all these are used by the plant, but simply that they have been in the soil solution and that the plant's protoplasm has been permeable to them. The plant is not selective. Selective absorption by the root is not possible, for if a diffusion gradient of a substance exists and the proloplast is permeable to it it will enter.

Certainly, if a salt molecule once inside the cells of the plant is taken out of the diffusional system, as by being "built into" another more complex molecule or is adsorbed, then the inward end of the gradient remains low. This will delay or postpone indefinitely the attainment of equilibrium between the plant's sap and the soil solution, and the gradient will not flatten out.

In these cases when the molecule is immobilized by adsorption, or taken out of the diffusional system by chemical action, more of the substances will appear in the plant, and an appearance of preferential or selective absorption is suggested.The point here is that as regards entry the living cells cannot pick and choose from amongst a mixture of molecules in solution to all of which it is permeable.

The term osmosis should be used only to describe the diffusion of the solvent of a system across a semi-permeable membrane. The term *dialysis* should be preferred in describing the passage of a solute. Once a solute molecule or ion, diffusing in the soil-water, has approached the root hair, it may proceed into the tissues by the same routes as a water molecule, and diffuse through the water of the colloidal cell walls or dialyse from cell to cell. The path for such a diffusing unit is along the water continuum. Eventually it will reach the water in the xylem.

There it will be carried up in solution by mass flow. Thus, water and dissolved salts are conducted to the shoot tissues for distribution amongst the mesophyll cells by diffusion. Transpiration takes off the water, and the salts are left for purposes of synthesis. The final result is very similar to concentration of a solution by evapouration. The first of these is what has been referred to as selective accumulation, and sometimes as "heaping up." In general, either of these two terms applies to all cases where the plant on analysis is found to contain more of a given ion or molecule than seems justified by the very weak concentration available to it in the culture medium.

Analysis of a nutrient solution after a plant capable of selective accumulation has been growing in it for some time suggests that selective absorption has been going on. The alteration in the composition of a water culture medium in which barley plants had been grown for twenty-four hours. The simplest case, and one to which the terms can only just be applied, occurs when an ion is used in building up a more complex molecule, which is part of the fabric of the plant.

SALTS INTAKE

The units of the precipitate are too large to be mobile. The actual gradient is in fact kept low at the inward end, and diffusion continues as long as precipitation within the cell is possible and there is a supply of the particle in solution outside.

Another case arises where the substance taken in is adsorbed on a surface inside the plant. Here, too, the particle is immobilized and taken out of the diffusional system. Again the net result is that diffusional equilibrium cannot be attained or is delayed.

Selective Accumulation in the Narrow Sense

The case where the term selective accumulation applies in the strict sense occurs when the concentration of any particular ion *in solution* inside the cell sap is higher than the concentration outside. The semi-permeable membrane is represented as separating the two solutions; the value of the ions are shown graphically. The substance continues to enter against the gradient.

The explanation of this phenomenon is not simple and not yet complete, but it is clear that in achieving this condition the plant expends energy. Accumulation against the gradient can only take place where energy release or respiration is active. That means there must be a supply of material which is usable as a "fuel" in respiration, and there must be oxygen present.

When these conditions suifor release of energy are satisfied certain ions are accumulated, and the cell sap contains *in solution* a higher concentration of these particles than does the external medium. Physically this is equivalent to making a body travel uphill. A number of factors in addition to food and oxygen supply affect the amount of solute forced in. Examples of these are temperature, transpiration rate, hydrogen ion concentration of the medium and the stage of growth of the plant. From the point of view of the grower of food this aspect of intake is important.

Plants capable of selective accumulation will have in their tissues a reserve of solute if at any time the supplying power of the soil or the intake capacity of the root should temporarily fall below requirements. Another facet of this question of selective accumulation is recognized when it is remembered that the soil often has a considerable power of holding many ions out of solution, either by adsorption or included in a but slightly soluble compound.

Potassium and phosphate are common examples. There may be abundant supplies of such substances in the soil, but only a very small proportion of them in solution and available to the plant.

Movement into the plant of the few ions actually in solution will cause others to be released from the reserves locked up in the soil. Selective accumulation in the plant, by maintaining the concentration of such an ion in the soil solution always at a minimum, causes a smoother release from the soil complex and builds up a reserve in the tissues readily available to meet any demand. In general, selective accumulation is to the plant what a cistern of water in the loft is to a house with a poor supply from the main pipes.

Interference between Particles

The second amendment to the simple diffusion theory of salt intake lies in the fact that certain pairs of ions affect each other in their entry into, or passage across, a membrane. If a high proportion of magnesium be present in the culture medium offered to a plant, injury is caused to the living cells of the root. This is due to a poisoning effect of magnesium. The locus of the poisoning is in the cell protoplast, probably at its surface, and acts possibly by altering the balance between the solid and liquid phases.

When, however, sufficient calcium is added to the medium the poisoning effect does not appear. Calcium is here said to act as a *depoisoner,* and the two elements are said to be antagonistic. The calcium does not reduce the amount of magnesium; it merely prevents its effect. The calcium probably acts by effecting repair of the protoplast or preventing the initial damage.

Elements Which Show Antagonism

A number of antagonistic pairs are known; for example Mg/Ca, S/Se, and Na/Ca. It is to be made clear that it is not the absolute amounts of the poisoner and its depoisoner which may be present that are important; it is the ratio between the two. In the reduction of the severe ill-effects caused by magnesium it has been shown that the percentage of injured leaves on a plant decreased as the Mg/Ca ratio increased from 1 to 50.

Experiments with a number of crop plants yielded data of the same type. It has been shown that the dry weight of tops of wheat plants were inversely proportional to the logarithm of the Mg/Ca ratio between the limits 1 and 50 of the ratio.

THE ELEMENTS WHICH PLANTS USE

THE fundamental mechanism of salt intake must be related to practice. The matter is important, for it is now recognized that estimation of the carbohydrates, fats, and proteins in a food is not a complete index of its value. The total ash (mineral) content and the proportion in which each component of it occurs is also very significant.

ESTIMATION OF MINERAL INTAKE

Consider first the methods by which the rate of entry and amount entering the plant of the various components may be followed. Two main lines of investigation may be recognized. On the one hand there are methods of chemical analysis, and on the other methods in which the reaction of a plant indicates the state of affairs.

Chemical Methods

Description of the detailed methods of the chemist are not within the province of this text. Briefly, the material of the plant body is first sampled, and a quantity dried in an oven to constant weight. The loss in weight indicates the water content. A more positive result may be obtained by distilling off the water with such a solvent as naphtha. The dry matter remaining from the oven test may be used for the estimation of the major food classes, but a weighed portion should be incmerated to provide the ash.

The ash is analysed by suimethods to reveal the proportions of each component present. There are available in the literature many examples of data obtained from this kind of work. Some of these carry the investigation to very fine limits.

Water Culture: Hydroponics

Another way in which purely chemical procedure may be employed lies in growing a plant in a medium of known composition, and after a period analysing the medium. This shows which components have been taken by the plant and what quantity of each.

Plants live and do well when their roots are immersed in a solution of mixed chemicals of appropriate strength and composition and so a technique of "water-culture" or "solution culture is commonly employed. So successful is the method that it has been adapted to commercial crop production, and is now called the science of hydroponics.

One disability of water-culture in experimental work is that the solutions are necessarily weak, and a component in demand by the plant is soon in deficient supply, thus upsetting the mineral balance of the solution. To obviate this the liquid in the culture jars should be changed often, or better still, the plant grown in a stream of fresh culture solution flowing steadily round the roots from a reservoir.

Sometimes the plants are grown in clean, well-washed sand, moistened with the known solution. In many investigations both the plants and the culture solution in which they grow are analysed. Plants grown in a solution of known composition (the changes in which are followed) are at the end of a period themselves analysed. A further refinement has been introduced.

Radio-active isotopes of certain elements are used in the formation of the mineral compounds to be offered to the plants. The presence of such isotopes may be readily detected at any time. Hence a radicle or ion with such an isotope is "tagged" or " labelled," and its path into and through the plant may be followed.

Biological Methods

Turning now to the methods by which the- reaction of the plant is used to reveal the state of affairs. The first of these is to use some special plant as a "living reagent." Certain strains of the common mould fungus *Aspergillus niger* can be used to estimate the copper content of a soil.

The colour of the spores of this plant are white when no copper is present. In the presence of 1 part of copper to 200 million parts of soil the spores are orange-yellow. The spores become light brown with 2 parts per 200 million, while, in concentrations up to 1 part in 16 millions, the spores produced are dark brown or black. The results are obtained in five days.

The same species of mould may also be used to test for magnesium deficiency. After five days n a nutrient medium complete except for magnesium this fungus made no growth. In exactly the same medium, but with 257 magnesium added (as $MgSO_27H_2O$) slight sterile growth appeared. In similar media, but with increasing amounts of magnesium added, the growth of the fungus improved and the number of spores increased.

Injection Methods

More usually it is one of the higher plants that is used in this type of test. When a crop appears "off colour" or shows some pathological condition, and some maladjustment of the mineral supply is suspected as the cause, a series of pure chemicals, each in separate solution, may be injected into it. Injection may be carried out in a number of ways. Merely painting a leaf or fruit with an appropriate solution of a salt will cause a local response.

The molecule in deficient supply diffuses from solution through the cuticle. The solution "runs back," and that solution which remedies the defect in the leaf is soon identified. The fluid may feed in through the stump of a cut leafstalk In the case of a woody tree an auger hole is bored in the trunk. Into this is led a supply pipe from a reservoir of appropriate solution.

After some time the transpiration pull draws the solution into the xylem, and so up into the leaves. In other cases pellets of appropriate chemicals may be forced into the wood of the tree trunk. There they dissolve in the sap and are carried to all parts of the plant. When appropriate chemicals are used a response is seen in the tree.

The Elements used by the Plant

By employing one or other of these various methods it has been established that plants must get from their environment, in quantity, carbon, hydrogen, oxygen, nitrogen, sulphur, phosphorus, potassium, calcium, magnesium, along with a smaller quantity of iron. The source of hydrogen and oxygen is the water supplied from the soil. Oxygen is also supplied from the air.

The green plant uses only carbon dioxide gas from the air as a source of carbon. The carbon of the organic matter in the soil is not used. It seems strange that an organism built entirely of carbon compounds should be limited to such a meagre source of the element, for the gas occurs in the atmosphere to the amount of only three hundredths of one per cent.

There is little doubt that green plants could use more carbon dioxide than occurs naturally in the air. Nitrogen, on the other hand, is abundant in the air, but this is not used as a source of supply by the green plant. Certainly, some of the lower plants such as bacteria and fungi do use gaseous nitrogen, but all of these have a source of energy other than light. Nitrogen for the normal green plant comes only from the inorganic nitrogen in the soil.

Seeming exceptions to this are those green plants which live in association with bacteria or other lowly types. All the other elements listed above, *e.g.* sulphur, phosphorus, potassium, calcium, magnesium, and iron, under normal conditions are derived wholly from the soil.

There is evidence from spraying and painting experiments that the shoot can absorb inorganic salts. As there is rarely any natural source of supply of these to the shoot, this may be ignored. Elements are never absorbed as such, they must always be offered to the plant in combination, They enter in either molecular or ionic form. They must be in the dissolved state, and therefore free to diffuse.

In the case of most elements all inorganic combinations are not equally suitable, and the plant appears to be strict in its requirements. All the evidence points to phosphorus being taken only as a phosphate radicle or ion, sulphur probably only as the sulphate, and so on. Elements which are required in quantity, so that a poor crop results if they are not present in substantial amounts, are called *essential elements* in the nutrition of plants. When one crop seems to demand a relatively larger amount of a particular element than other crops do, that element is called the *dominant* ingredient for that crop.

Thus in crops grown for leaf (cabbages, kale, etc.) nitrogen is required in more quantity than is necessary for, mangolds, which seem rather to require a higher proportion of potassium. Nitrogen is regarded as the dominant ingredient for kale, while potassium is the dominant for mangolds. Interest next focuses on certain elements such as arsenic and copper, which at one time were regarded as entirely deleterious to the plant's health, and indeed could cause death.

These substances were classified as *toxic elements.* Many of our weed-killers or herbicides have developed from studies on these plant poisons. Other elements which appeared to have no effect when present were referred to as *passive* or *inert.* For a time the elements available to the plant were classified into essential, toxic, and passive.

Elements Required in Small Amounts

Early in the 'twenties of this century evidence came forward that this classification was invalid. Largely through refinements in methods of water culture and improvements in the purification of salts, it was discovered that very small quantities of some elements, previously classified as poisonous or passive, are absolutely necessary for the growth and development of plants. The difference between these elements and the essential elements, so far as this discussion has gone, lies only in the amounts required of each.

Minor or Trace Elements

The name to be used for this class of elements essential for the health of the plant, but required only in very minute quantities, has not yet been established. Minor, micro- or trace elements have been proposed. A further term, micronutrient element, has been suggested, but has not been widely adopted, for there is no proof that they all enter into nutrition in the strict sense. There is something to be said for and against the use of each of the four terms, and here "minor" will be adopted.

The use of the term "rare" element in this connection is definitely to be deprecated. Many of the elements previously classified as poisonous or toxic, when supplied in sufficiently high dilution, are now found to be essential. These are the minor elements. By small quantity or high dilution is meant from about o-oi milligrams up to a few tenths of a milligram per litre of nutrient solution, or round about one part per million.

Deficiency of supply of such minor elements as boron, manganese, and copper cause pathological symptoms in the plant, and "disease" associated with over- or under-supply of nutrients will be considered later under pathology. Only the functional role of the various elements will be discussed now.

Quantity or Balance in Nutrient Supply

The elements absorbed through the roots and required for the complete nutrition of the plant (and eventually for the animal) form a series, graduated on the basis of the quantity required. Quantity in this regard should not be used in an absolute sense but rather in the sense of ratio. Too much of one may impose deficiency of another. Imbalance of the components of the nutrient solution presented to the plant is rather the better view.

This concept of imbalance applies to all the elements required by the plant for full health, and not to minor elements only. It must be stressed that a statement about deficiency of this or that element in the soil often does not necessarily imply complete absence of the substance. Rather it implies an insufficient proportion of available soluble compounds of the particular substance in question.

Any compound which is insoluble or otherwise prevented from diffusing is not available. Soil colloids adsorb or bind potassium, and the amount of this bound material available to the plant depends, not on the total amount present in the soil, but on how far the colloidal material is saturated with it.

Factors Affecting Ash Content

The total ash content and the relative proportion of each element found in the plant is conditioned by a number of circumstances. The first of these is the nature of the plant concerned. Two plants of the same kind growing in the same soil may be indistinguishable in appearance, yet differ in ash content.

These two belong to different physiological strains or varieties of the plant. Secondly, one and the same plant growing continuously under non-varying conditions may differ in ash content and composition at different stages of growth and in different organs.

Again, the nature of the supply to the plant from the soil or culture medium affects the nature of the ash. It is of interest that different plants offered either soluble or insoluble supplies of the same element show differences in ability to use them. Furthermore, addition of supplementary manures may not increase the ash in the plant.

If a fully fertile soil is used as a culture medium, the addition to it of soluble phosphate will not increase the amount of phosphorus in the tissues of the plant growing thereon. The water content of the soil, too, or the molecular concentration of salts in a liquid medium affects the total ash percentage and the proportions of each component in the tissues. The effect of species must be interpreted always with regard to the fact that there may be genetical strains within the species.

When this is so, the plant selector or breeder can intervene and develop specialized races. At the moment the more gross plant unit, the species, will suffice for consideration. An even larger unit may suffice for more extreme differences. Clovers and their allies (legumes in general) have a higher total ash content than the grasses growing alongside them. So, too, the clovers tend to have a higher proportion of lime to phosphorus, while the grasses have higher phosphorus to lime content.

The Ca/P_2O_5 ratio therefore tends to move towards calcium in the legume and towards phosphorus in the grasses. Within the grasses the kinds which normally constitute bog hay (a mixture of natural grasses found on poor, rather acid soils) have a lower total ash content than the better grasses found in the sown fields. A complete study of this subject is due to workers at the Waite Research Institute in South Australia, and the data discussed here will be taken from their work. Experimental plants were grown in large, glazed earthenware pots.

The soil used was a well-known type whose behaviour is understood. A gravel mulch some quarter of an inch on the surface of the soil reduced evapourational losses of water to the minimum. By adding weighed quantities of water the soil was maintained through the period of the experiment at 60 per cent, of its total water-holding capacity. The course of transpiration was therefore easily followed from the results of periodical weighings of the pots.

The plants were kept under experiment right through their life cycle up to the commencement of flowering, and hence loss of nutrients, ash, etc., from pollen-shedding, withering, and severance of leaves and flowers was obviated. Total yields are given in column i. It is seen that in this respect species differ very considerably. These figures are in conformity with general observations made in the field. The three cultigens, *Lolium subulatum* (a ryegrass), *Trifolium subterraneum* (a clover), and barley (a cereal) easily outyield the "natural"-plants included in the experiment.

They appear to yield very poorly, but in their native habitat they have a value because they produce what they do when little feed is available from other sources. The high "soluble ash" content of these species further enhances their value. Only difficulties in the establishment of these plants would seem to prevent their cultivation in suiareas. The position of *Erodium* in the is of interest.

The difference in ash content between the clover and the three members of the grass family *(Danthonia,* barley, and *Lolium subulatum)* follow very much the general position already stated. While remembering that these figures relate to plants as a whole at the flowering stage, it seems that the low yield of dry matter given by *Erodium* is important because it has a high content of each of nitrogen, phosphate, lime, and magnesia.

The clover is runner-up in lime content, and takes a high place in regard to the other substances. The high phosphate content of *Erodium* and *Atriplex vesicarium* is interesting, and the value of these plants on this account is recognized by Australian stock-masters, who favour the plants for the beneficial effect they exert when eaten by stock.

The Use cf Insoluble Compounds by Certain Plants

That the mineral content is not evenly distributed throughout the plant is shown by the below. It is interesting to see here how phosphate and nitrogen migrate into the seed, while potash and lime remain more in the vegetative portion of the plant. It is essential that a substance be in solution before it can enter into an intact plant root. An element forming part of a compound which is insoluble in the soil water is not available to the plant so long as it remains unaltered chemically.

Plant roots excrete certain ions which react with compounds in the soil, and so render available the components of these insoluble materials. If a plant is grown in such a way that its roots come into actual contact with the polished surface of a slab of marble, it will be found after some time that the roots have dissolved away the calcium carbonate wherever contact has been made. The polished surface will show an etched tracery of the root.

The root, probably by excretion of carbon dioxide, dissolves the carbonate. This gives the results obtained from growing three very different plant species in complete water culture. Two different sets of solution were used. In one the phosphate was supplied as tri-calcic-phosphate (insoluble), and the other as mono-calcic-phosphate (soluble). Briefly, the ryegrass nor the clover can make use of insoluble phosphate to any great extent while erodium can.

In the case of clover, the yield of dry matter fell in response to the paucity of phosphate, but in such yield as there was the phosphate content was maintained. The ryegrass, when offered only insoluble phosphate, largely maintained its dry matter yield, but the percentage of phosphate dropped very considerably. The erodium largely maintained its yield and ash content in both the nutrient media. These three plants exemplify three general types.

- Plants which when offered only insoluble phosphate produce a reduced yield of normal herbage *(e.g.* clover).
- Plants which while not using insoluble phosphate keep up the yield of dry matter, and thus produce phosphate-deficient herbage *(e.g.* ryegrass).
- Plants which, presumably by their root action, can use insoluble phosphate, and so keep up yield of dry matter and mineral content *(e.g.* erodium).

Plants of the ryegrass type in a soil containing insoluble phosphate only, produce herbage which, while apparently feeding the stock well, is not supplying at least one essential element of the diet. The erodium type of plant, if it were capable of producing a better total yield, would be of importance because it would provide a good crop of normal herbage on soil rich in insoluble phosphate but deficient in an available (soluble) form.

Erodium is not the only kind of plant which falls into this class, and it would appear that these have a use on land deficient in available nutrients, but having a reasonable content of these in non-available forms.

Stage of Growth as Affecting Ash Content

The stage of growth of the plant is another factor affecting the percentage of mineral nutrients in the tissues. In experiments "harvests" were made at six developmental stages, namely:

1. Commencement of tillering;
2. Active tillering;
3. Advanced tillering;
4. Plants in full flower;
5. Plants with grain at "milk" stage;
6. Plants fully mature.

These show that at each succeeding stage there is a progressive fall in the *percentage present* of all nutrients derived from the soil except silica. With ryegrass and clover the curves are of slightly different shape, in that the major portion of intake comes rather later. These differences between plants is summarized for the before-and-after flowering periods.

The remainder of the data all go to show that the percentage of nitrogen and percentage of ash in the tissues fall progressively with age. These changes in the chemical composition of the tissues are due to progressive changes in the proportions of the kinds of tissue elements present in the plant.

With advancing age and development there is an increase in the amount of purely structural elements, such as sclerenchyma, xylem, and so on. These consist only of cell walls; they contain no living contents rich in nutrients. As the cells modify, the contents are broken down and translocated to younger areas.

THE UTILIZATION OF DIFFERENT COMPOUNDS OF ELEMENTS

Nitrogen

Let us now turn to the special characteristics of each element in connection with intake. It was believed at one time that the plant used nitrogen only as the nitrate ion NO_3, and that ammonia and other forms had to be converted to this radicle in the soil before entry or utilization was possible. It is now known that in the field ammonia is used satisfactorily by some plants. Whether any particular kind of plant can make use of ammonia or not would seem to depend on whether it can "step up" this form of nitrogen quickly to higher combinations such as an amide. Apparently ammonia cannot be stored as such in the cells.

The rate of intake of ammonia seems to be faster than the intake of nitrate—at least in such plants as the cotton seedling, oats, and buckwheat. Whether ammonia or nitrate is the more suiform seems to be governed by at least two factors, the nature of the plant and the reaction of the soil. An acid soil favours the use of nitrate.

The successful use of nitrite (NO_2) by plants has been reported, but is of little agricultural interest. The chief use of nitrogen in the plant is in the formation of compounds such as amino-acids, which enter into the molecular structure of the proteins. Since proteins are the basis of protoplasm, reduced nitrogen supply soon shows itself in reduction of protoplasm formation. This puts a restriction on practically all functions, and particularly on new cell formation, so that growth slows and may cease.

Sulphur

Sulphur is used by the plant only when offered in the form of a sulphate (SO_4); most other forms are toxic or not usable. The sulphur enters into certain amino-acids which are essential in the formation of many proteins. Many plants can maintain apparently normal growth and appearance while suffering from partial lack of sulphur. The tissues are, however, deficient in the element. A peculiar case of sulphur deficiency occurs in some grassland areas only in spring when growth of the grass is very rapid.

The sulphur intake at this time is not sufficient to supply what is necessary for normal protein building, but the plant goes on growing, using such sulphur as it has very economically, or improvising in some other way. A mass of herbage deficient in sulphur is produced. Later in the year, when growth slows down, the sulphur supply is sufficient, so that the maladjustment corrects itself, and normal protein formation follows.

Break in Wool

As wool protein contains much sulphur, sheep grazing in such an area during the spring develop a peculiar characteristic of the wool. When a lock of the wool is sharply pulled, each fibre snaps at the same point, and the broken tuft shows a clean break instead of the irregular break obtained with normal wool. When wool fibres of this type are seen under the microscope an identical part of each fibre is seen to be of reduced diameter.

This is the weak point, and is where the break takes place. Working from known development rates for wool fibre, it appears that the weak part is produced while the sheep is feeding on the spring flush of grass, which is deficient in sulphur. The condition is remedied by feeding during the deficiency period a "lick" of hydrolized wool, which supplies organic sulphur.

Selenium

It will be convenient to discuss -one of the non-essential elements here as it is intimately related to sulphur. This is selenium, which is of special interest because it is often absorbed by plants in a quantity sufficient to render them toxic to animals while themselves remaining unaffected.

The element is comparatively rare, but traces of its compounds in the soil result in herbage, grain, etc., producing all the symptoms of the dreaded alkali disease when fed to animals. Man, too, is subject to selenium poisoning.

There is no doubt that plants "heap up" selenium, and that the accumulation is due in large part to the metal replacing the sulphur in certain of the amino-acids of the protein molecule. The process of "heaping up" need not go very far for the plant to become toxic to animals. Rats die when fed on perfectly normal-appearing plants containing 2 or 3 parts per miJlion of selenium. On the other hand, "heaping-up" must go very far before the plants show symptoms.

Fifteen parts per million in the plant produces chlorosis, and with greater dosage the leaves and roots develop a pink colouration. Fortunately selenium and sulphur are antagonistic. Addition of sulphur to the soil substantially reduces the amount of selenium subsequently found in the tissues of the plant. By applications of gypsum, selenium-containing soil can be made to produce crops which are non-toxic to animals.

Selenium-tolerant Plants

Some plants not only tolerate this element so toxic to animals but actually require it for successful growth. *Astragalus racemosus* and *A. pattersonia* are reported as thriving in a medium containing 27 parts of selenium per million. This is about ten times the amount tolerated by wheat or buckwheat. Furthermore, in these plants addition of sulphur to the soil, while resulting in considerable reduction of the selenium content, does not bring it down to a level likely to be tolerated by grazing animals.

Plants which occur naturally on seleniferous soils and nowhere else are now becoming known. It is expected that delimitation of seleniferous areas by mapping the occurrence and distribution of these "indicator" plants will be comparatively simple, and so eliminate much chemical work on soil analysis.

Selenium as an Insecticide

An interesting development of these discoveries regarding selenium is that crops can be protected from biting or sucking insects if selenium is added to the soil in which the plants grow. Insects die after feeding on the seleniferous juice. This can be made use of without detriment to man or his animals only when the harvested produce is not consumed as food, *e.g.* timber and fibres, or where it is food but not protein, *e.g.* sugar.

A slight exception to this is seen in cacao where the very young plant is subject to attacks from insect pests. The juvenile plantlets are supplied for a short period with selenium, after which they are transplanted to fresh untreated land. By the time the plants are adult and bearing a crop the small amount of this element which protected the small plant is dispersed through a large plant, and the small percentage then present is not toxic in the product.

Phosphorus

Phosphorus is of fundamental importance in the economy of the plant, involved as it is in protein formation, as well as in respiration and other functions. The utilization of nitrogen by the plant is interfered with if phosphorus is in deficient supply. This may be due directly to insufficiency of phosphate for the formation of protein. There is evidence, however, that the action lies deeper, for when phosphates are deficient the enzyme responsible for the initial steps in nitrate utilization is either absent or inoperative.

Lack of phosphate seems, then, to "bottleneck "the whole of the synthesis in the plant. Mono- and di-saccharides "bank-up" amino and amide nitrogen accumulate, and eventually cell-division slows down and ceases.

The slowing down of cell-division may be a reflex of deficiency of necessary phosphates for protoplasm formation, but it is also known that When phosphorus supply is inadequate the fibres which "pull" the split chromosomes apart in mitosis are not formed. A concomitant of this whole imbalance of cell function is cessation of all phenomena requiring active cell-division, *e.g.* growth of the axis and branching.

This is particularly apparent in cereals where the result is that tillering is depressed. The' converse of these effects is seen when phosphate is supplied liberally, for there is then more active branching of the root and shoot, while the plant tends to flower and seed not only earlier but more profusely.

Calcium

The predominant role of calcium in agriculture is not wholly botanical, but arises from the use of lime as a soil ameliorant and as a reducer of soil acidity. The element, however, is taken into the plant in quite substantial quantities, probably as the metallic ion. In the tissues it neutralizes acids arising from many metabolic activities, and crystals of the calcium salts of oxalic and other organic acids commonly appear in cells. One of the principal roles of calcium in the plant seems to be involved in some fine balance with boron, and possibly other elements.

What the relationship is still remains obscure. Certainly deficiencies of both boron and calcium are very similar in the symptoms they produce. This is especially so in the failure of nornytLetevelopment in meristematic regions. Calcium is found in great quantities in leaves, and does not migrate much from an old organ to a young one as the plant develops. Calcium is an essential ingredient of the structure of the cell wall, or at least of the middle lamella.

The Calcium/Phosphorus Ratio

The importance in the diet of animals of adequate quantities of calcium and phosphorus individually, and of the ratio of these to each other, needs no stressing here. The data offered in the general account of salt intake should be again examined from this aspect of the quality of food.

Potassium

Potassium in plant nutrition offers something of a paradox in that commonly there is a considerable quantity of potash in the soil with only a small proportion soluble, while in the plant practically the whole of the potassium remains soluble in the cell sap. When a plant tissue is mashed up the potassium may be easily leached out.

Despite its appearance in this simple form potassium has fundamental roles to perform. Potassium is concerned with the synthesis of carbohydrates, and probably with their translocation. This connection between potassium and carbohydrates is clearly seen in crops which normally accumulate large reserves of starch or sugar.

These crops when grown in a deficiency of sunlight produce less reserve food (lower crop yield), but when generously supplied with potash the production is greatly increased. It would appear that potassium can "replace" sunlight. Potassium is moved out of adult tissues to areas of meristematic activity.

Potash confers on plants some degree of resistance to various diseases. The clovers and their allies, though not "carbohydrate" plants, gain greater benefit from increased potash supplies than do grasses, and hence in pastures with a grass/clover sward addition of potash tends to strengthen clovers rather than the grasses.

Magnesium

Magnesium is an essential ingredient for the formation of chlorophyll, and indeed is the only metallic constituent of this all-important green pigment of plants. As the plant passes to final maturity and the leaves change colour, magnesium is largely transferred from the vegetative body to the seed. In the plant another of its functions is connected with its use along with phosphate in the formation of those special proteins which form the cell nucleus (the nucleo-proteins).

Magnesium, too, appears to be concerned in oil metabolism. The amount of magnesium required by the plant is not large and, when too great quantity is available in the soil functional disturbances develop in the plant.

Iron

Iron is not an essential element in the sense that the term has so far been used, for only a little is required. Neither is it regarded as a minor element, but occupies an intermediate position in the classification. Iron seems to take no structural part in the plant, yet its presence is essential for the formation of chlorophyll. Though necessary for the formation of chlorophyll, it does not enter into the molecule of that substance. Iron is not mobile inside the plant.

This is easily shown by first growing plants for a time in a solution supplying iron when the shoot is fully green. If the experimental plants are then transferred to a culture medium deficient in iron, the new leaves produced in further growth are white due to a lack of iron in them inhibiting formation of chlorophyll. Total absence of iron from soils is not so common as total or nearly total non-availability. Thus it is that many soils rich in iron produce plants showing all the symptoms of iron deficiency.

THE MINOR ELEMENTS IN PLANT NUTRITION

Boron

Of all the minor elements probably none has received so much attention as boron. It is deficient in quantity in many soils all over the world. Symptoms of disfunction following from apparent boron deficiency seem to be connected with calcium supply, and show fundamentally in reduction or stoppage of activity in the meristematic regions.

Further, plants showing symptoms of boron deficiency are found usually in calcareous soils or those which have been heavily limed. Boron does not seem to be necessary to animals, so that the chief interest in this minor nutrient resides in the pathological condition it may set up in the plant.

Manganese

Manganese deficiency symptoms, too, are most commonly found in plants growing on soils well supplied with lime. Except at high dilution manganese is toxic to plants, yet when presented in too low a concentration metabolic upsets result. These are definitely connected with lowered efficiency of the respiratory function. Indeed manganese seems to be necessary for the whole oxidation/reduction mechanism.

This has been shown in connection with the use of ammonia by plants. Plants offered a nitrogen supply in the form of ammonia compounds are benefited by extremely small additions of manganese to the nutrient medium. This is especially so when the oxygen supply is limited. Additions of manganese seem to benefit the legumes more than other cultivated plants. A sufficiency of manganese in the food is necessary for animals.

Copper

Copper has long been known as an element toxic to plants even when present in what may be considered high dilution. One part in 10 million of water is toxic to fresh-water algae and copper sulphate is commonly used in the elimination of water weeds from tanks, etc. This element is receiving more and more attention as a minor element.

Its increasing importance arises from the fact that copper deficiency causes pathological symptoms in animals as well as in plants. In the higher plants a modicum of copper is essential, and many soils, often those of a peaty character, fail to provide the very small amount necessary. The addition of to of one part per million in water culture, or about *1 per cent* to 3 Ib. of copper per acre, seems to eliminate all trouble in this respect.

The element is often supplied in the form of the sulphate or pyrites: the amount of either of these to be sown depends on their copper content. The mechanism through which the copper acts is not known. That plants deficient in copper are often lacking in chlorophyll is not without importance, for copper deficiency in the diet of infants is often the cause of some of the anaemias that they suffer from.

The red pigment of blood and the green pigment of the plant are related in very many respects. Where copper is in too great supply it has been shown that iron acts as a "depoisoner." Deficiency of zinc, too, causes pathological symptoms in the plant. It is thought that the curative action of some fungicidal zinc sprays is more connected with the action of the metal on the physiology of the host that its direct effect on the disease organism.

Cobalt

Soils deficient in cobalt produce herbage of normal appearance, but the animals fed on the produce pine and die.

The addition of about 2 lb. cobalt chloride per acre eliminates the trouble.

Molybdenum

Molybdenum was fairly early shown to be necessary for the nutrition of many fungi, but it is now thought that higher plants use it. Certain soils, typically those lying on the Lower Lias formation in south-east England, contain about 20 to 100 parts per million of this element, and the herbage produced on them may contain about 2 to 3 parts per million. When grazing on such a pasture ruminant animals scour, pine, and eventually die. Such pastures are known as "teart."

Clovers seem to absorb more molybdenum than grasses, hence increase of these valuable plants intensifies the bad effect. It would seem that the "availability" of molybdenum is controlled by the acidity of the soil. In neutral or alkaline soils the availability is high; in acid conditions low. Addition of lime to the soil does not render teart pastures healthy, but rather the reverse. Also, the clovers being encouraged by the lime become more prolific, and because of their high molybdenum content they raise the content of the herbage as a whole.

There is some evidence that copper may be antagonistic to molybdenum and be useful in treating affected animals or in improving teart pastures. Other elements affect the plant and through the plant, animals. Enough has been said to elucidate the principles involved, and for fuller details of each case reference should be made to the very voluminous literature which has already grown up.

Chapter 10

Reproduction of Plants

The way plants reproduce is a totally fascinating process. Out of the more than 300,000 different kinds of plants, more than half are seed plants. Seed plants make their own seeds from which new plants grow. Other ways plants can make new plants are from spores, rhizomes, bulbs, tubers, corms, cuttings, grafts, and buds. Seeds are made by flowers in some plants and by cones in other plants.

Flower parts contain specialized cells, called ovules, which have the job of making the seeds. Different kinds of flowers have many different parts. Basically, most plants--trees, vegetables, and even grass--sprout flowers. Some of these flowers are beautiful and colourful while others are so tiny one can hardly see them, but all plants produce new plants.

Botanists classify the reproduction of plants as ones with perfect and imperfect flowers. Perfect flowers are the ones that contain both the male and female parts. Imperfect flowers are the ones that contain only male or only female parts. So, of course, these different flowers have to find different ways to propagate themselves.

The stamen is the male part of a flower and has an anther on a stalk. The anther's job is to produce the pollen. Then there is the pistil, or female part. The pistil contains the flower's ovary, the style, and the stigma. Inside the ovary are the ovules. Each ovule contains an egg cell. So when an egg cell joins with a pollen cell, a seed may be produced.

The ovary is at the base of the plant. The stigma is at the top, and between them is the style. So, the pollen from the anther has to land on the stigma. When that happens, the pollen germinates and makes its way down into the ovary. Inside the ovary are the ovules, which contain egg cells. When an egg cell joins with a pollen cell, a seed may be produced. Then the ovary becomes a fruit. So a fruit is a mature ovary holding the seeds.

In fruits, like the pears, the pear's core is the seedpod of the pear tree. Acorns are the seedpods of oak trees. And the tomato is the seedpod of the tomato plant.

A flower garden is alive with insects. They are all after the sweet liquid called nectar, which the flowers produce, near the bottom of the pistil. Flowers produce the nectar to draw the insects to them. Insects--like butterflies, wasps, and bees--come to get the nectar. Some, like butterflies and wasps, use the nectar for food; some, like bees, make it into honey and store it in their hives.

If you look closely at the body of a butterfly or bee, you will notice it is covered with tiny hairs. As they land on the flower to collect the nectar, some of the pollen is brushed on these hairs. Then the insect carries the pollen to the next flower, and pollination takes place.

Some plants, like corn, or flowers of an oak tree, rely on the wind to spread their pollen. So these kinds of plants don't produce pretty flowers since they don't need to attract insects to come and pollinate them.

Some plants need long tongues to pollinate them. Their flowers are usually tubular, like the trumpet creeper vine. These kinds of flowers are best pollinated by hummingbirds or by moths that have long tongues. And flowers that open at night do so because they need the specialized pollination provided by certain moths. Even the colour of plants makes a difference. Birds, like the hummingbird, usually pollinate plants that are red. Night insects pollinate white and pale yellow flowers, which are seen better at night.

Trees that have needle-like leaves are called conifers. Conifers produce cones instead of flowers. Under each scale of a cone is a part that produces either pollen or an ovule. But a cone cannot produce both. The very small cones produce the pollen in the spring. The wind carries the pollen to the larger cone, which has the ovule producing part. Then, the pollen producing cones fall off while the fertilized ovules grow larger. Eventually, they open and release the seeds, and some of the seeds sprout. Some plants, like tulips, grow from bulbs. Bulbs produce plants very easily by reproducing underground.

Plants can be grown without seeds by cuttings. A cutting can be from any part of the fully-grown plant, and when it is placed in soil, it may produce a new plant. You can increase your own houseplants by this method, and here is the way to do it. With a sharp knife, cut a piece of stem from the houseplant you want to use. Pull off any leaves or flowers from the stem's lower part. Stick the stem in some clean, moist sand. Cover the flowerpot with a clear, plastic bag. Place in a well-lit place. After a few weeks, lift the stem out of the sand. Are the roots beginning to grow? Now you can plant it in a pot with potting soil. Keep the soil moist, and soon, the new plant will begin to grow.

When cuttings from trees and other woody stems are attached to another plant, this is known as grafting. Grafts are another way of growing plants without using seeds.

The remarkable regenerative capacity of plant-members is largely made use of for the artificial propagation of plants. A branch removed from a parent-plant will, under appropriate conditions, develop roots, and so constitute a new plant; this is the theory of propagation by "cuttings." A portion of a root will similarly develop one or more shoots, and thus give rise to a new plant. An isolated leaf will, in many cases, produce a shoot and a root, that is, a new plant; it is in this way that new begonias, for instance, are propagated.

The production of plants from leaves occurs also in nature, as, for instance, in certain so-called "viviparous" plants, of which *Bryophyllum calycinum* (Crassulaceae) and many ferns are examples. But it is in the mosses, of all plants, that the capacity for vegetative propagation is most widely diffused. Any part of a moss, whether it be the stem, the leaves, the rhizoids, or the sporogonium, is capable, under appropriate conditions, of giving rise to filamentous protonema, on which new moss-plants are then developed as lateral buds.

The simplest case of vegetative multiplication is afforded by unicellular plants. When the cell which constitutes the body of the plant has attained its limit of size it gives rise to two either by division or gemmation; the two cells then grow, and at the same time become separated from each other, so that eventually two new distinct individuals are produced, each of which precisely resembles the original organism.

A good example of this is to be found in the germination of the yeast plant. This mode of multiplication is simply the result of the ordinary processes of growth. All plant-cells grow and divide at some time or other of their life; but whereas in multicellular plant the products of division remain coherent, and add to the number of the cells of which the plant consists, in a unicellular plant they separate and constitute new individuals. In more highly organized plants vegetative propagation may be effected by the separation of the different parts of the body from each other, each such part developing the missing members and thus constituting a new individual.

This takes place spontaneously in rhizomatous plants, in which the main stem gradually dies away from behind forwards; the lateral branches thus become isolated and constitute new individuals. In a large number of plants provision is made for vegetative propagation by the development of more or less highly specialized organs. In lichens, for instance, there are the *soredia,* which are minute buds of the thallus containing both algal and fungal elements; these are set free on the surface in large numbers, and each grows into a thallus.

In the Characeae there are the *bulbils* or "starch-stars" of *Chara stelligera,* which are underground nodes, and the branches with naked base and the proembryonic branches found by Pringsheim on old nodes of *Chara fragilis.* In the mosses small tuberous bulbils frequently occur, on the rhizoids, and in many instances stalked fusiform or lenticular multicellular bodies containing chlorophyll, termed *gemmae,* are produced on the shoots, either in the axils sof the leaves or in special receptacles at the summit of the stem. Gemmae of this kind are produced in vast numbers in *Marchantia* and *Lunularia* among the liverworts.

Similar gemmae are also produced by the prothallia of ferns. In some ferns the buds borne on the leaves or in their axils become swollen and filled with nutritive materials, constituting *bulbils* which fall off and give rise to new plants.

This conversion of buds into bulbils, which subserve vegetative multiplication, occurs also occasionally among Phanerogams, as for instance in *Lilium bulbiferum,* species of *Poa, Polygonum viviparum.* But many other adaptations of the same kind -occur among Phanerogams. Bulbous plants, for instance, produce each year at least one bulb or corm from which a new plant is produced in the succeeding year. In the potato, tubers are developed from subterranean shoots, each of which in the following year gives rise to a new individual. In the dahlia, *Thladiantha dubia.,* tuberous swellings are found on the roots, from each of which a new individual may spring.

TRUE REPRODUCTION

This is effected by cells formed by the proper reproductive organs. These cells are of two principal kinds. There are, first, those cells each of which is capable of developing by itself into a new organism: these are the *asexual* reproductive cells, known generally as *spores.* Secondly, there are the cells which are incapable of independent germination; it is not until these cells have fused together in pairs that a new organism can be developed: these are the *sexual* reproductive cells or *gametes.* In some exceptional cases the normal mode of reproduction, sexual or asexual, does not take place: instead, the new organism is developed vegetatively from the parent.

When sexual reproduction is suppressed the case is one of *apogamy;* when asexual reproduction by spores is suppressed the case is one of *apospory. Asexual Reproduction. - Reproduction* by means of some kind of spore (using the term in its widest sense, so as to include all asexually produced reproductive cells) is common to nearly all families of plants; it is wanting in certain Algae (Conjugatae, Fucaceae, Characeae), and in certain fungi.

The structure of a spore is essentially this: it consists of a nucleated mass of protoplasm, enclosing starch or oil as reserve nutritive material, usually invested by a cell-wall. In those cases in which the spore is capable of germinating immediately on its development the cell-wall is a single delicate membrane consisting of cellulose; but in those cases in which the spore may or must pass through a period of quiescence before germination the wall becomes thickened and may consist of two layers, an inner, the *endospore,* which is delicate and consists of cellulose, and an outer, the *exospore,* which is thick and rigid, frequently darkly coloured and beset externally with spines or bosses, and which consists of cutin.

In some few cases among the fungi, multicellular or septate spores are produced; these approximate somewhat to the gemmae as highly specialized organs for vegetative propagation. In some cases, particularly among the algae, and also in some fungi (Peronosporeae, Saprolegnieae, Chytridiaceae, and the Myxomycetes), spores are produced which are usually destitute of any cell-wall, and are further peculiar in that they are motile, and are therefore termed *zoospores;* they move sometimes in an amoeboid manner by the protrusion of pseudopodia, but more frequently they are provided with one, two, or many delicate vibratile protoplasmic filaments, termed *cilia,* by the lashing of which the spore is propelled through the water.

The zoospore eventually comes to rest, withdraws its cilia, surrounds itself with a cell-wall, and then germinates.

In the simplest case a single spore is developed from the cell of the unicellular plant, the protoplasm of which surrounds itself with the characteristic thick wall. This occurs only in plants of low organization such as the Schizophyta. In other cases the contents of the cell undergo division, each portion of the protoplasm constituting a spore. Examples of this are afforded, among unicellular plants, by yeast and the Protococcaceae; and in multicellular plants by the Pandorineae, Confervaceae, Ulvaceae, &c., where any cell of the body may produce spores.

In such cases the spore-producing cell may be regarded as a rudimentary reproductive organ of the nature of a *sporangium*. In more highly organized plants special organs are differentiated for the production of spores. In the majority of cases the special organ is a sporangium, that is, a capsule in the interior of which the spores are developed; but in many fungi the spores are formed by abstriction from an organ termed a *sporophore*. In the Thallophyta the sporangium is commonly a single cell. In the Bryophyta it is a multicellular capsule. In the Pteridophyta the sporangium is multicellular, but simple in structure, and this is true also of the Phanerogams.

It is important to note that in all the Bryophyta and in some of the Pteridophyta (most of the Filicinae, all existing Equisetinae, and the Lycopodiaceae and Psilotaceae) there is but one kind of sporangium and spore, the plants being *homosporous* or isosporous, whereas the rest of the Pteridophyta (Hydropterideae, Selaginellaceae) and the Phanerogams are *heterosporous*, having sporangia of two kinds; some produce one or a few large spores (*megaspores*), and are hence termed *megasporangia*, while others give rise to a larger number of small spores (*microspores*) and are hence termed *microsporangia*.

In the Phanerogams the two kinds of sporangia have received special names: the megasporangium, which produces as a rule only one mature spore (*embryo-sac*), is termed the *ovule*; the microsporangiurn, which produces a large number of microspores (*pollen-grains*), is termed the *pollen-sac*. The development of spores, except in the simpler Thallophyta, is more or less restricted to definite parts of the body.

Thus in the Red Algae (Florideae) there are the organs known as *stichidia, nemathecia.* In the fungi the number and variety of such organs is very great; they may be described generally as simple and compound *sporophores.* In the higher plants the organs are less various.

In the Bryophyta the production of spores is restricted to the sporogonium. In the vascular plants (Pteridophyta, Phanerogams) the development of sporangia, speaking generally, is confined to the leaves. In most ferns the sporangiferous leaves (*sporophylls*) do not differ in appearance from the foliage leaves; but in other Pteridophyta (Equisetaceae, Marsiliaceae, some species of *Lycopodium* and *Selaginella*) they present considerable adaptation, and notably in the Phanerogams.

In the Phanerogams the specialization is so great that the sporophylls have received special names; those which bear the microsporangia (pollen-sacs) are termed the *stamens,* and those which bear the megasporangia (ovules) are termed the *carpels.* The sporophylls are usually aggregated together on a short stem, forming a shoot that constitutes a flower.

Many terms are employed to indicate the nature of the various kinds of spores, especially among the fungi, but the endless varieties of asexual reproductive cells may be grouped under two heads - *Gonidia, Spores proper.* The distinction between these two kinds of asexual reproductive cells is as follows.

The *gonidium* is a reproductive cell that gives rise, on germination, to an organism resembling the parent. For instance, among the algae, the "zoospore" of *Vaucheria* develops into a Vaucheria-plant. There is thus a close connexion between vegetative multiplication and multiplication by means of gonida. The production of gonida is entirely limited to the Thallophyta, and is especially marked in the fungi, though the nature of all the many kinds of reproductive cells formed in this group has not yet been fully investigated. It is, however, wanting in certain algae and fungi.

The *spore proper* is a reproductive cell that as a rule gives rise, on germination, to an organism unlike that which produced it. For instance, the spore of a fern when it germinates gives rise, not to a fern-plant, but to a prothallium. The apparent exceptions to this rule occur only among the Thallophyta, and are explained below in the section on *Life-history*. The true spore is developed, usually in a sporangium, after a process of division which presents certain features that call for special notice.

Observation of the process of division of the nucleus (*karyokinesis*) in plants generally has shown that the linin-reticulum of the resting nucleus breaks up into a definite number of segments, the *chromosomes*, each of which bears a series of minute bodies, the *chromatin-disks* or *chromomeres*, consisting largely of a substance termed *chromatin*. In the ordinary *homotype* divisions of the nuclei the characteristic number of chromosomes is always observable: but when the spore-mother-cells are being formed the number of chromosomes is reduced to one-half.

This, if the number of chromosomes of the parent plant be expressed as *2x* the number in the spore will be *x*. To take a concrete case: it has been observed by Guignard and others that in the early divisions taking place in the developing anther and ovule of the lily the number of chromosomes is 24; whereas in the later divisions which give rise to the pollen-mother-cells in the one case and to the mothercell of the embryo-sac in the other, the number of chromosomes is only 12.

Thus the development of a spore is always preceded by a *reducing-* or *heterotypedivision*, a process now more generally termed *meiosis*. The reduced number of chromosomes in the nucleus of the spore-mother-cell persists in the spore, and in all the cells of the organism to which the spore may give rise. It should be explained that cells, to which the name "spore" has also been applied, are formed as the result of a sexual act: such are *zygospores*, *oospores*, and some *carpospores*. But these cells differ from spores proper not only in their mode of origin but also in that their nuclei contain the full double number (2x) of chromosomes; hence they may be distinguished as *diplospores*. *Sexual Reproduction*. - *Sexual* reproduction involves the development of sexual organs (*gametangia*) and sexual cells (*gametes*).

When the organism is unicellular, as in the lower Green Algae (e.g. Protococcaceae, Conjugatae), the cell becomes a sexual organ and its whole protoplasm gives rise to one or more sexual cells: in the higher forms certain parts of the body are specialized as sexual organs. In many of the lower plants the organs present no external distinction of sex it is impossible to distinguish between the male and female organs, although it cannot be doubted that the essential physiological difference exists; consequently the organs are merely described as gametangia. The gap between these plants and those with differentiated sexual organs is, however, bridged over by intermediate forms.

When the sexual organs are more or less obviously differentiated into male and female, they present considerable variety of form in different groups of plants, and accordingly bear different names. Thus the male organ is a *pollinodium* in most of the fungi, a *spermogonium* in others (certain Ascomycetes, Uredineae); in all other plants it is an *antheridium*. Similarly the female organ is an *oogonium* in various Thallophyta (Green and Brown Algae: Oomycetous Fungi); a *procarp* in the Red Algae; an *archicarp* in certain Ascomycetous Fungi and in the Uredineae; an *archegonium* in all the higher plants.

It is generally the case that the protoplasm of the sexual organ is differentiated into one or more sexual cells. Thus the gametangium usually gives rise to cells which, as they are externally similar, are termed *isogametes* or simply *gametes*. Certain forms of the male organ, the spermogonium and the antheridium, give rise to male cells which are termed *spermatic* when they are nonciliate, *spermatozoids* when they are ciliated and free-swimming. Again, the female organs termed oogonia and archegonia produce one or more female cells called *oospheres*.

But there are important exceptions to this rule. Thus the protoplasm is not differentiated into cells in the gametangium of the Mucorinae; in the male organ (pollinodium), of fungi generally; and in the female organ (procarp) of the Red Algae and (archicarp) of the Ascomycetes and Uredineae.

The immediate product of the fusion of cells, or of undifferentiated protoplasm, derived from sexual organs of opposite sex may be generally termed the *zygote;* but it is not always of the same kind. Thus when two isogametes, or the undifferentiated contents of two gametangia, fuse together, the process is designated *conjugation,* and the product is usually a single cell termed *zygospore.* When an oosphere fuses with a male cell, or with the undifferentiated contents of a male organ, the process is *fertilization,* and the product is a single cell termed *oospore.*

When, finally, a female organ with undifferentiated contents receives a male cell, the process again is fertilization; here the product is not a single cell, but a fructification termed *cystocarp* (Red Algae), or *ascocarp* (Ascomycetes) or *aecidium* (Uredineae), containing many spores (*carpospores).* As a consequence of the diversity in the sexual organs and cells, in the details of the sexual act, and in the product of it, several modes of the sexual process have to be distinguished, which may be conveniently summarized as follows:

1. *Isogamy:* the sexual process consists in the fusion of either two similar sexual cells (*isogametes*), or two similar sexual organs (*gametangia)* : it is termed conjugation, and the product is a *zygospore.* Its varieties are
    a. Gametes ciliated and free-swimming (*planogametes*), set free into the water where they meet and fuse: lower Green Algae (Protococcaceae, Pandorineae, most Siphonaceae and Confervaceae); some Brown Algae (Phaeosporeae):
    b. Gametangia fuse in pairs, and a gamete is differentiated in each: the gametes of each pair

fuse, but are not set free and are not ciliated (the Conjugate Green Algae): or, no gametes are differentiated, the undifferentiated contents of the gametangia fusing (Mucorinae among the Fungi).

2. *Oogamy:* male and female organs distinct: the protoplasm, of the female organ is differentiated into one or (rarely) more oospheres which usually remain enclosed in the female organ: the contents of the male organ are usually differentiated into one or more male cells: the process is fertilization, the product is an *oospore.* The sexual organs are unicellular (or coenocytic as in certain Siphonaceous Green Algae and in the Oorycetous Fungi); the female organ is an *oogonium.*

    a. The male organ is an *antheridium* giving rise to one or more free-swimming ciliated spermatozoids:

        - The oogonium contains a single oosphere which is fertilized *in situ:* higher Green Algae (*Volvox, Vaucheria, Oedogonium, Coleochaete,* Characeae); some Brown Algae (*Tilopteris);* among the Fungi, *Monoblepharis,* the only fungus known to have spermatozoids:
        - The oogonium produces a single oosphere which is extruded and is fertilized in the water: *Dictyota* and some Fucaceae (Brown Algae):
        - The oogonium contains several oospheres which are fertilized *in situ: Sphaeroplea* (Siphonaceous Green Alga):
        - The oogonium produces more than one oosphere (2-8) which are extruded and are fertilized in the water: certain Brown Algae (*Pelvetia, Ascophyllum, Fucus):* The male organ is a *pollinodium* which applies itself closely to the oogonium: the amorphous male cell is not ciliated and is not set free:

- The oogonium contains a single oosphere which is fertilized *in situ:* Peronosporaceae (Oomycetes):
- The oogonium contains several oospheres; Saprolegniaceae: but it is debated whether or not fertilization actually takes place.

b. The male and female organs are (as a rule) multicellular; the male organ is an *antheridium,* the female an *archegonium:* the archegonium always contains a single oosphere which is fertilized *in situ.*

   - The male cell is a free-swimming ciliated spermatozoid: the antheridium produces more than one (usually very many) spermatozoids, each of which is developed in a single cell: all Bryophyta and Pteridophyta: the only Phanerogams in which spermatozoids have been observed are the gymnospermous species *Ginkgo biloba, Cycas revoluta, Zamia integrifolia.* The male cell is amorphous and passes directly from the pollen-tube into the oosphere (*siphonogamy*).

   It must be explained that in the angiospermous Phanerogams, the male and female organs are so reduced that each is represented by only a single cell: the male, by the *generative cell,* formed in the pollen-grain, which usually divides into two male cells: the female, by the oosphere. The gradual reduction can be traced through the Gymnosperms.

   Attention may here be drawn to the fact that, in several cases, the second male cell has been seen to enter the embryo-sac from the pollen-tube, and its nucleus to fuse with the definitive nucleus (endosperm-nucleus) or with one of the polar nuclei.

3. *Carpogamy:* the sexual organs are (as a rule) differentiated into male and female: the protoplasm

of the unicellular or multicellular female organ (archicarp, procarp) is never differentiated into an oosphere: in many cases definite male cells, *spermatia*, are produced and are set free, but they are not ciliated, and frequently have a cell-wall: the process is *fertilization:* the product is a fructification derived essentially from the female organ con taining several (sometimes very many) spores (*carpospores*): characteristic of the Red Algae and of the Ascomycetous Fungi.

a. There are definite male cells (*spermatia*): The female organ is a *procarp*, consisting of an elongated, closed, receptive filament, the *trichogyne*, and of a basal fertile portion, the *carpogonium:* on fertilization the latter grows and gives rise directly or indirectly to a *cysto-* carp: the spermatia are each formed in a unicellular antheridium and have no cell-wall at first: they fuse with the tip of the trichogyne: Red Algae (Rhodophyceae or Florideae): The female organ (*archicarp*) resembles the preceding: in fertilization the fertile portion (*ascogonium*) develops into an *ascocarp* containing one or more asci (sporangia) each containing usually eight *ascospores:* the spermatia are formed by abstriction from the filaments (*sterigmata*) lining special receptacles, the *spermogonia*, which are the male organs: certain Ascomycetous Fungi (e.g. Laboulbeniaceae, some Lichen-Fungi, *Polystigma*).

b. There are no definite male cells: the more or less distinct male and female organs come into contact, and their undifferentiated contents fuse: the product is an *ascocarp:* The male and female organs are obviously different: the female organ is an *ascogonium*, the male a *pollinodium: e.g. Pyronema, Sphaerotheca* (Ascomycetes): The male and female organs are quite similar: *e.g. Eremascus, Dipodascus* (Ascomycetes).

It may be explained that carpogamy is the expression of sexual degeneration. When the sexual organs are quite similar, they have reverted to the condition of gametangia. Still further reduction is observable in other Ascomycetes in which one of the sexual organs, presumably the male, is either much reduced or is altogether wanting. Again in the rusts (Uredineae), there are spermatia, but they are functionless. In the highest Fungi, the Autobasidiomycetes, no sexual organs have been discovered.

DETAILS OF THE SEXUAL ACT

It has been already stated that the sexual act consists in the fusion of two masses of protoplasm, commonly cells, derived from two organs of opposite sex: but this is only the first stage in the process.

The second stage is the fusion of the nuclei, which usually follows quickly upon the fusion of the cells; but nuclear fusion may be postponed so that the two sexual nuclei may be observed in the zygote, as "conjugate" nuclei, and even in the cells of the organism developed from the zygote (e.g. Uredineae). The result of nuclear fusion is that the nucleus of the zygote contains the double number of chromosomes - that is, if the number of chromosomes in each of the fusing sexual nuclei be x, the number in the nucleus of the zygote will be 2X.

Moreover, this double number persists in all the cells of the organism developed from the zygote, until it is reduced to one-half by meiosis preceding either the development of the spores, or, less commonly, the development of the sexual cells.

But there is yet a third stage, which consists in the temporary fusion of the chromosomes belonging to the two sexual nuclei. This always takes place as a preliminary to meiosis; it may be in the germinating zygote, or after many generations of cells have been formed from it. At the onset of meiosis the (2x) chromosomes are seen to be double, one of each pair having been derived from the male and the female cell respectively: the chromosomes of each pair then fuse so that their chromomeres unite along their length, constituting the *pseudo-chromosomes.*

The paired chromosomes separate and eventually go to form the two daughter-nuclei, one to each, which thus have half the original number of chromosomes. The daughter-nuclei at once divide homotypically, retaining the reduced number of chromosomes to form the four nuclei of a tetrad of spores (more rarely, *e.g. Fucus,* of sexual cells).

Life-history

It will have been gathered from the foregoing sections that plants generally are capable of both sexual and asexual reproduction; and, further, that in different stages of their life-history they possess the diploid (2x) number of chromosomes in their nuclei, or the haploid (x) number.

It may be at once stated that, in all plants in which sexual reproduction and true meiotic spore-formation exist, these two modes of reproduction are restricted to distinct forms of the plant; the sexual form bears only the sexual organs and is haploid; the asexual form only produces spores and is diploid. Hence all such plants are to this extent *polymorphic - that* is, the plant assumes these two forms in the course of its life-history.

When, as in many Thallophyta, one or other of these forms can reproduce itself by means of gonidia, additional forms may be introduced into the lifehistory, which becomes the more complicated the more pronounced the polymorphism. The most straight forward life-histories are those presented by the Bryophyta and the Pteridophyta, where there are but the two forms, the sexual and the asexual. In the life-history of a moss, the plant itself bears only sexual organs: it is the sexual form, and is distinguished as the *gametophyte.*

The zygote (oospore) formed in the sexual act develops into an organism, the *sporogonium,* which is entirely asexual, producing only spores: it is distinguished as the *sporophyte.* When these spores germinate, they give rise to moss-plants.

Thus the two forms, the sexual and the asexual, regularly alternate with each other - that is, the life-history presents that simple form of polymorphism which is known as *alternation of generations*. Similarly, in the life-history of a fern, there is a regular alternation of a sporophyte, which is the fern-plant itself, with a gametophyte, which is the fern-prothallium.

It is pointed out in the preceding section that, as the result of the sexual act, the nucleus of the zygote contains twice as many chromosomes as those of the fusing sexual cells. This 2X number of chromosomes persists throughout all the cellgenerations derived from the zygote, that is, in the cells constituting the sporophyte, up to the time that it begins to produce spores, when meiosis takes place.

Again, the cellgenerations derived from the spore, that is, the cells constituting the gametophyte, all have the reduced *x* number of chromosomes in their nuclei up to the sexual act. Hence the sporophyte may also be designated the *diplophyte* and the gametophyte the *haplophyte* (Strasburger): in other words, the sporophyte is the *pre-meiotic,* the gametophyte the post-meiotic generation. Twice in its life-history the plant is represented by a single cell: by the spore and by the zygote. The turningpoints in the life-history, the transitions from the one generation to the other, are meiosis, the sexual act.

The course of the life-history in Phanerogams and in those Thallophyta which have been adequately investigated is essentially the same as that of the Bryophyta and of the Pteridophyta as described above, though it is less easy to trace on account of the peculiar relation of the two generations to each other in the Phanerogams and on account of various irregularities that present themselves in the Thallophyta.

In the Phanerogams, as in the Pteridophyta, the preponderating generation is the sporophyte, the plant itself. Inasmuch as they are heterosporous, the gametophyte is represented by a male and a female organism or prothallium, both rudimentary. The male prothallium consists of the few cells formed by the germinating pollen-grain (microspore); and though it is quite independent, since the microspores are shed, it grows parasitically in the tissues upon which the microspore has been deposited in pollination.

The female prothallium may consist of many cells with well-developed archegonia, as in the Gymnosperms, or of only a few cells with the female organ reduced to the oosphere, as in the Angiosperms. In either case it is the product of the germination of a megaspore (embryo-sac) which is not shed from its sporangium (ovule): hence it never becomes an independent plant, and was long regarded as merely a part of the sporophyte until its true nature was ascertained, chiefly by the researches of Hofmeister, who first explained the alternation of generations in plants. This intimate and persistent connexion between the two generations affords the explanation of the characteristic features of the Phanerogams, the seed and the flower.

The ovule containing the embryo-sac, which eventually contains the embryo, persists as the seed - a structure that is distinctive of Phanerogams, which have, in fact, on this account been also termed Spermatophyta.

With regard to the flower it is, like the cone of an Equisetum or a Lyco podium, a shoot adapted to the production of spores. But it is something more than this: for whereas in Equisetum or Lycopodium the function of the cone comes to an end when the spores are shed, the flower of the Phanerogam has still various functions to perform after the maturation of the spores. It is the seat of the process of *pollination - that* is, the bringing of the pollen-grain by one of various agencies into such a position that a part (the pollen-tube) of the male prothallium developed from it may reach and fertilize the oosphere in the embryo-sac.

Thus the flower of Phanerogams is a reproductive shoot adapted not only for spore-production, but also for pollination, for fertilization, and for the consequences of fertilization, the production of seed and fruit. However, in spite of these complications, it is possible to determine accurately the limits of the two generations by the observation of the nuclei. The meiosis preceding the formation of the spores marks the beginning of the (haploid) gametophyte, male and female; and the sexual act marks that of the (diploid) sporophyte.

The difficult task of elucidating the life-histories of the Thallophyta has been successfully performed in certain cases by the application of the method of chromosome-counting,. with the result that alternation of generations has been found to be of general occurrence.

To begin with the Algae. In. the Dictyotaceae (Brown Algae) there are two very similar forms in the life-history, the one bearing asexual reproductive organs (tetrasporangia), the other bearing sexual organs (oogonia. and antheridia). It has been shown that the former is undoubtedly the sporophyte and the latter the gametophyte, since the nuclei of the former contain 3 2 chromosomes, and those of the latter 16. Meiosis takes place in the mother-cell of the tetraspores, which, on germination, give rise to the sexual form.

Quite a different life-history has been traced in *Fucus*, another Brown Alga. Here no spores are produced: there is but one form in the life-history, the *Fucus* plant, which bears sexual organs and has, on that account, been regarded as a gametophyte. The investigation of the nuclei has, however, shown (Farmer) that the Fucus-plant is actually diploid, that it is, in fact, a sporophyte; but since there is no spore-formation, meiosis immediately precedes the development of the sexual cells, which alone represent the gametophyte.

Similarly, two types of life-history have been discovered in the Red Algae. In *Polysiphonia violacea*, a species in which. the tetraspores and the sexual organs are borne by similar but. distinct individuals, it has been ascertained (Yamanouchi) that, as in *Dictyota*, meiosis takes place in the mother-cell of the tetraspores, so that the nuclei of these spores, as also those of the sexual plants to which they give rise, contain 20 chromosomes: and further, that the nuclei of the carpospores (diplospores) produced in the cystocarp as the result of fertilization,. contain 40 chromosomes, as do also those of the asexual plant. to which the carpospores give rise.

Hence the sporophyte is represented by the cystocarp and the resulting tetrasporangiate plants: the gametophyte, by the sexual plants. Though it is the rule in the Red Algae that the tetrasporangia and the sexual organs are borne on distinct individuals, yet cases are known in which both kinds of reproductive organs arse borne upon the same plant; and to those the above conclusions obviously cannot apply. They have yet to be investigated.

The second type of life-history has been traced in *Nemalion*. Here there is no tetrasporangiate form, consequently meiosis. takes place at a different stage in the life-history. It has. been observed that the nuclei of the sexual plant contain 8 chromosomes; those of the gonimoblast-filaments of the developing cystocarp contain r6, whilst those of the carpospores contain 8: hence meiosis takes place in the carposporangia.

Here the plant is the gametophyte; the sporophyte is only represented by the cystocarp. The carpospores here are true spores (haplospores). Among the Green Algae, *Coleochaete* is the only form that has been fully investigated (Allen). Here meiosis takes place in the germinating oospore: consequently the plant is the gametophyte, and the sporophyte is represented only by the oospore, so that the life-history resembles that of *Nemalion*.

It is probable that this conclusion is generally true of the whole group; at any rate of those forms (Desmids, *spirogyra, Oedogonium, Chara*) which have been more or less investigated. Turning to the Fungi, somewhat similar results have been obtained in the few forms that have been studied from this point of view. In the sexual Ascomycetes it appears (Harper) that meiosis takes place in the ascocarp just before the development of the spores, so that the life-history essentially resembles that of *Nemalion*.

Again, in certain Uredineae, having an aecidium-stage and a teleutospore-stage, which is apparently a sexual process has been observed (Blackman, Christman) which is described in the section on *Abnormalities of Reproduction*, and the life-history is as follows. The sexual act having taken place, a row of aecidiospores is developed in the aecidium, each of which contains two conjugate nuclei derived from the sexual nuclei. The mycelium developed from the aecidiospore, as well as the uredospores and the teleutospores that it bears, shows two conjugate nuclei.

When, however, the teleutospore is about to germinate, the two nuclei fuse (thus completing the sexual act) and meiosis takes place. As a result the promycelium developed from the teleutospore, and the sporidia that it produces, are uninucleate: so are also the mycelium developed from the sporidium, and the female organs (archicarps) borne upon it. Hence the limits of the sporophyte are the aecidiospore and the teleutospore: those of the gametophyte, the teleutospore and the aecidiospore.

Similar observations have been made upon other Uredineae with a more contracted life-history. *Phragmidium Potentillaecanadensis* is a rust that has no aecidium-stage: consequently the primary uredospores are borne by the mycelium produced on infection of the host by a sporidium.

It has been observed that the sporogenous hyphae fuse in pairs, suggesting a sexual act; then the primary uredospores are developed in rows from the fused pairs of hyphae which thus behave as sexual organs (archicarps), and each such uredospore contains two conjugate nuclei. Although the research has not been carried beyond this point, it may be inferred that in this case, as in the preceding, nuclear fusion and meiosis take place in the teleutospore. Here the sporophyte is represented by the uredo-form. Finally, in some of the fungi in which no sexual organs have yet been discovered, this method of investigation has made it probable that some kind of sexual act takes place nevertheless.

Thus in the Uredine *Puccinia malvacearum,* which has only teleutosporeand sporidium-stages, it has been observed that the formation of the teleutospores is preceded by a binucleate condition of the hyphae. The same idea is suggested by the binucleate basidia of the Basidiomycetes, which correspond to the teleutospores of the Uredineae. The life-histories sketched in the preceding paragraphs show that one of the complexities met with in the Thallophyta is that meiosis does not always take place at the same point in the life-history. In the higher plants the incidence of meiosis is generally, though not absolutely, constant: it may be stated as a rule that in the Bryophyta, Pteridophyta and Phanerogams it takes place in the spore-mother-cells. In the Thallophyta this rule does not hold.

In some of them, it is true, meiosis immediately precedes, as in the higher plants, the formation of certain spores, the tetraspores (Dictyotaceae, *Polysiphonia*), the teleutospores (Uredineae): but in others it immediately precedes the development of the sexual organs (Fucaceae), or follows more or less directly upon the sexual act (Green Algae, *Nemalion*, Ascomycetes).

The life-history of most Thallophyta is further complicated by the capacity of the gametophyte of the sporophyte to reproduce themselves by cells termed gonidia, a capacity that is wholly lacking in the higher plants. The karyology of gonidia has not yet been sufficiently investigated: but when, as in the Green Algae and the Oomycetous Fungi, the gonidia are developed by and reproduce the gametophyte, it may be inferred that they, like the gametophyte, are haploid.

One case, at any rate, of the reproduction of the sporophyte by gonidia is fully known, that of the Uredineae just described, in which the uredoform, which is a phase of the sporophyte, is reproduced by the uredo-spores which are binucleate, that is diploid, and may be distinguished as *diplogonidia*. In any case the result is that whereas in the higher plants each of the alternating generations occurs but once in the life-history, in these Thallophyta the lifehistory may include a succession of gametophytic or of sporophytic forms, This is, in fact, a distinguishing feature of the group.

The higher plants present a *regular* alternation of generations: whereas, in the Thallophyta, though they probably all present some kind of alternation of generations.

Sufficient information has been given in the preceding pages to render possible the consideration of the *origin* of alternation of generations. To begin quite at the beginning, it may be assumed that the primitive form of reproduction was purely vegetative, merely division of the unicellular organism when it had attained the limits of its own growth.

Following on this came reproduction by a gonidium: that is, the protoplasm of the cell, at the end of its vegetative life, became quiescent, surrounded itself with a proper wall, or was set free as a motile ciliated cell, having in some unexplained way become capable of originating a new course of life (*rejuvenescence*) on germination..

Then, as can be well traced in the Brown and Green Algae, these primitive reproductive cells (gonidia) began to fuse in pairs: in other words, they gradually became sexual. This stage can still be observed in some of these Algae (e.g. *Ulothrix, Rctocarpus*) where the zoospores (gonidia) may either germinate independently, or fuse in pairs to form a zygote. Gradually the sexuality of these cells became more pronounced: losing the capacity for independent germination, they acquired the external characters of more or less differentiated sexual cells, and the gametangia producing them developed into male and female sexual organs.

But this advancing sexual differentiation did not necessarily deprive the plant of the primitive mode of propagation: the sexual organism still retained the faculty of reproduction by gonidia. The loss of this faculty only came with higher development: it is entirely wanting in some of the higher Thallophyta (e.g. Fucaceae, Characeae), and in all plants above them in the evolutionary series.

With the introduction of the sexual act, a new kind of reproductive cell made its appearance, the zygote. This cell, as. already explained, differs from other kinds of spores and from the sexual cells, in that its necleus is diploid; and with it the sporophyte (diplophyte) was introduced into the life-history.

In some plants (e.g. Green Algae) the zygote is all that there is to represent the sporophyte, giving rise, or germination and after meiosis, to one or more spores. Passing to the Bryophyta, in the simpler forms (e.g. *Riccia*), the zygote develops into a multicellular capsule (sporogonium); and in the higher forms into a more elaborate sporogonium, producing many spores. In the Pteridophyta and the Phanerogams, the zygote gives rise to the highly developed sporophytic plant.

Thus the evolution of the sporophyte can be traced from the unicellular zygote, gradually increasing in bulk and in independence until it becomes the equal of the gametophyte (e.g. in *Dictyota* and *Polysiphonia*), and eventually far surpasses it (Pteridophyta, Phanerogams).

Moreover, the increase in size was attended by the gradual limitation of spore-production to certain parts only, the rest of the tissues being vegetative, assuming the form of stems, leaves, &c. These facts have been formulated in the theory of "progressive sterilization", which states that the sporophytic form of the higher plants has been evolved from the simple, entirely fertile, sporophyte of the lower, by the gradually increasing development of the sterile vegetative tissue at the expense of the sporogenous, accompanied by increase in total bulk and in morphological and histological differentiation.

In connexion with the study of the evolution of the sporophyte, the question arose as to its morphological significance; whether it is to be regarded as a modified form of the gametophyte, or as an altogether new form intercalated in the life-history: in other words, whether the alternation is "homologous" or "antithetic." In certain plants there is a succession of forms which are undoubtedly homologous: for instance, in *Coleochaete* where a succession of individuals without sexual organs is produced by zoospores (gonidia).

The main fact that has been established is that the sporophyte, from the simple zygote of the Thallophyta to the spore-bearing plant of the Phanerogams, is characterized by its diploid nuclei; that it is a diplophyte, in contrast to the haplophytic gametophyte.

Were these nuclear characters absolutely universal, there could be no question but that the sporophyte is an altogether new antithetic form, and not an homologous generation. But certain exceptions to the rule have been detected, which are described under *Abnormalities of Reproduction:* at present it will suffice to say that such things as a diploid gametophyte and a haploid sporophyte have been observed in certain ferns.

It can only be inferred that alternation of generations is not absolutely dependent upon the periodic halving in meiosis and the subsequent doubling by a sexual act, of the number of chromosomes in the nuclei, though the two sets of phenomena usually coincide. It must not, however, be overlooked that these exceptional cases occur in plants presenting an abnormal life-history: the fact remains that where there is both normal spore-formation with meiosis, and a subsequent sexual act, the haploid form is the gametophyte, the diploid the sporophyte.

But the actual observation of a haploid sporophyte and of a diploid gametophyte makes it clear that however generally useful the nuclear characters may be in the distinction of sporophyte and gametophyte, they do not afford an absolute criterion, and therefore their value in determining homologies is debatable.

Abnormalities of Reproduction

In what may be regarded as the type of normal life-history, the transition from the one generation to the other is marked by definite processes: there is the meiotic development of spores by the sporophyte, and the sexual production of a zygote, or something analogous to it, by the gametophyte. In the preceding pages that the transition may, in certain cases, be effected in other ways, which may be regarded as abnormal, though they are constant enough in the plants in which they occur, in fact as manifestations of reproductive degeneration.

In the first place, the sporophyte may be developed either after an abnormal sexual act, or without any preceding sexual act at all, a condition known as *apogamy*. In the second, the gametophyte may be developed otherwise than from a post meiotic spore, a condition known as *apospory*. Apogamy. - The cases to be considered under this head may be arranged in two groups:

1. *Pseudapogarny: sexual act abnormal*: *The* following abnormalities have been observed:
   a. *Fusion of two female organs*: observed (Christman) in certain Uredineae (*Caeoma nitens, Phragmidium speciosum, Uromyces Caladii*) where adjacent archicarps fuse: male cells (spermatia) are present but functionless.
   b. *Fusion between nuclei of the same female organ*: observed in the ascogonium of certain Ascomycetes, *Humaria granulate* (Blackman), where there is no male organ; *Lachnea stercorea* (Fraser), where the male organ (pollinodium) is present but is apparently functionless.
   c. *Fusion of a female organ with an adjucent tissue-cell*: observed (Blackman) in the archicarp of some Uredineae (*Phragmidium violaceum, Uromyces Poae, Puccinia Poarum)* : male cells (spermatia) present but functionless.
   d. There is no female organ: fusion takes place between two adjacent tissue-cells of the gametophyte; the sporophyte is developed from diploid cells thus produced, but there is no proper zygote as there is in *a, b* and *c* : observed (Farmer) in the prothallium of certain ferns (*Lastraea pseudo-mas,* var. *polydactyla)* : male organs (and sometimes female) present but functionless. Another such case is that of *Humaria rutilans* (Ascomvcete), in which nuclear fusion has been observed (Fraser) in hyphae of the hypothecium: the asci are developed from these hyphae, and in them meiosis takes place; there are no sexual organs.

2. *Eu-apogamy: no kind of sexual act.* The gametophyte is haploid:
   a. The sporophyte is developed from the unfertilized oosphere: no such case of *true parthenogenesis* has yet been observed.

      The sporophyte is developed vegetatively from the gametophyte and is haploid: observed in the prothallia of certain ferns, *Lastraea pseudo-mas,* var. *cristata-apospora(Farmer* and Digby), and *Nephrodium molle* (Yamanouchi).
   b. *The gametophyte is diploid*: The sporophyte is developed from the diploid oosphere: observed in some Pteridophyta, viz. certain ferns (Farmer), *Athyrium Filix-foemina,* var. *clarissima, Scolopendrium vulgare,* var. *crispum-Drummondae,* and *Marsilia* (Strasburger); also in some Phanerogams, viz. Compositae (*Taraxacum,* Murbeck; *Antennaria alpina,* Juel; sp. of *Hieracium* (Rosenberg): Rosaceae: Ranunculaceae (*Thalictrum purpurascens,* Overton).

Many other ferns are known to be apogamous, but they are not included here because the details of their nuclear structure have not been investigated. APosPoRY. - The known modes of apospory may be arranged as follows: *Pseudapospory: a spore is formed but without meiosis, so that it is* diploid - observed only in heterosporous plants, viz. certain species of *Marsilia (e.g. Marsilia Drummondii*) where the megaspore has a diploid nucleus (32 chromosomes) and the resulting prothallium and female organs are also diploid (Strasburger); and in various Phanerogams, some Compositae (*Taraxacum* and *Antennaria alpina,* Juel), some Rosaceae (*Eu-Alchemilla,* Strasburger), and occasionally in *Thalictrum purpurascens* (Overton), where the megaspore (embryosac) is diploid; in some species of *Hieracium* it has been found (Rosenberg) that adventitious diploid embryo-sacs are developed in the nucellus: these plants are also apogamous.

Eu-apospory: no spore is formed - of this there are two varieties:

a. With meiosis: this occurs in some Thallophyta which form no spores; the sporophyte of the Fucaceae bears no spores, consequently meiosis takes place in the developing sexual organs; the Conjugate Green Algae also have no spores, meiosis taking place in the germinating zygospore which develops directly into the sexual plant.

b. Without meiosis: the gametophyte is developed upon the sporophyte by budding; that is, spore-reproduction is replaced by a vegetative process: for instance, in mosses it has been found possible to induce the development of protonema, the first stage of the gametophyte, from tissuecells of the sporogonium: similarly, in certain ferns (varieties of *A thyrium Filix-foemina, Scolopendrium vulgare, Lastraea pseudo-mas, Polystichum angulare,* and in the species *Pteris aquilina* and *Asplenium dimorphum*), the gametophyte (prothallium) is developed by budding on the leaf of the sporophyte, and in some of these cases it has been ascertained that the gametophyte so developed has the same number (2x) of chromosomes in its nuclei as the sporophyte that bears it - that is, it is diploid.

Apospory has been found to be frequently associated with apogamy; in fact, in the absence of meiosis, this association would appear to be inevitable.

Combined Apospory and Apogamy. - Instances have been given of the occurrence of both apospory and apogamy in the same life-history; but in all of them there is a regular succession of sporophyte and gametophyte. The cases now to be considered are those in which one or other of the generations gives rise directly to its like, sporophyte to sporophyte, gametophyte to gametophyte, the normally intervening generation being omitted.

It is possible to conceive of this abbreviation of the life-history taking place in various ways. Thus, a sporophyte might be developed from a haploid spore instead of a gametophyte as is the normal case, but this has not been observed: again, a sporophyte might be developed from a diploid spore (as distinguished from a zygote or a diploid oosphere), a possibility that is to some extent realised in the life-history of some Uredineae in which successive forms of the polymorphic sporophyte are developed from diplogonidia.

Similarly a gametophyte might be developed from a fertilized or an unfertilized female cell: the latter possibility is to some extent realised in those Algae (e.g. *Ulothrix, Ectocarpus*) in which the sexual cells (isogametes), if they fail to conjugate, germinate independently as gonidia, giving rise to gametophytes.

The more familiar mode is that of vegetative budding. When a "viviparous" fern or Phanerogam reproduces itself by a bud or a bulbil, both spore-formation and the sexual act are passed over: sporophyte springs from sporophyte.

Remarkable cases of this have been observed in certain Phanerogams (*Coelebogyne ilicifolia, Funkia ovata, Nothoscordum fragrans, Citrus,* sp. of *Euonymus, Opuntia vulgaris*) in the ovule of which adventitious embryos are formed by budding from cells of the nucellus: with the exception of *Coelebogyne,* it appears that this only takes place after the oosphere has been fertilized. In other plants it is the gametophyte that reproduces itself by means of gemmae or bulbils, as commonly in the Bryophyta, the prothallia of ferns.

The abnormalities described are all traceable to reproductive degeneration; the final result of which is that true reproduction is replaced more or less completely by vegetative propagation. It may be inquired whether degeneration may have proceeded so far in any plant of sufficiently high organization to present spore-formation, or sexual reproduction, or both, as to cause the plant to reproduce itself entirely and exclusively by the vegetative method.

The only such case that suggests itself is that of *Caulerpa* and possibly some other Siphonaceous Green Algae. In this plant no special reproductive organs have yet been discovered, and it certainly reproduces itself by the breaking off of portions of the body which become complete plants: but it is quite possible that reproductive organs may yet be discovered.

Physiology of Reproduction

The reproductive capacity of plants, as of animals, depends upon the fact that the whole or part of the protoplasm of the individual can develop into one or more new organisms in one or other of several possible ways. Thus, in the case of unicellular plants, the whole of the protoplasm of the parent gives rise, whether by simple division or otherwise, to one or more new plants.

Reproduction necessarily closes the life of the individual: here, as August Weismann long ago pointed out, there is no natural death, for the whole of the protoplasm of the parent continues to live in the progeny. In multicellular plants, on the contrary, the reproductive function is mainly discharged by certain parts of the body, the reproductive organs, the remainder of the body being essentially vegetative - that is, concerned with the maintenance of the individual. In these plants it is only a part of the protoplasm that continues to live in their progeny; the remainder, the vegetative part, eventually dies.

It is therefore possible to distinguish in them, on the one hand, the essentially reproductive protoplasm, which may be designated by Weismann's term *germ-plasm,* though without necessarily adopting all that his use of it implies, and the essentially vegetative, mortal protoplasm, the *somatoplasm,* on the other. In the unicellular plant no such distinction can be drawn, for the whole .of the protoplasm is concerned in reproduction.

But even in the most highly organized multicellular plant this distinction is not absolute: for, as already explained, plants can, in general, be propagated by the isolation of almost any part of the body, that is vegetatively, and this implies the presence of germ-plasm elsewhere than in the special reproductive organs.

If the attempt be made to distinguish between the organs of vegetative propagation and those of true reproduction, the nearest approach would be the statement that the former contain both germ-plasm and somatoplasm, whereas the latter, or at least the reproductive cells, consist entirely of germ-plasm.

The question now arises as to the exact seat of the germ-plasm, and the answer is to be looked for in the results of the numerous researches into the structure and development of the reproductive cells that form so large a part of the biological work of recent years. The various facts to prove that the nucleus plays the leading part in the reproductive processes of whatever kind: the general conclusion is justified that no reproductive cell can develop into a new organism if deprived of its nucleus.

It may be inferred that the nucleus either actually contains the germ-plasm, or that it controls and directs the activities of the germ-plasm present in the cell. It is not improbable that both these inferences may be true. At any rate there is no sufficient ground for excluding the cooperation of the cytoplasm, especially of that part of it distinguished as *kinoplasm,* in the reproductive processes.

Pursuing the ascertained facts with regard to the nucleus, it is established that the part of it especially concerned is the linin-network which consists of the chromosomes. The behaviour, as already described, of the chromosomes in the various reproductive processes has led to the conclusion that the hereditary characters of the parent or parents are transmitted in and by them to the progeny: that they constitute, in fact, the material basis of heredity. They can hardly, however, be regarded as the ultimate structural units, for the simple reason that their number is far too small in relation to the transmissible characters.

It has been suggested that the chromomeres are the units, but the number of these would seem to be hardly sufficient. It seems necessary to fall back upon hypothetical ultimate particles, as suggested by Darwin, de Vries and Weismann, which may be generally termed *pangens.* The chromomeres may be regarded as aggregates of such particles, the "ids" of Weismann.

The foregoing considerations make it possible to attempt an explanation of the various reproductive processes.

VEGETATIVE PROPAGATION

It is easily intelligible that the two individuals produced by the division of a unicellular plant should resemble the parent and each other; for, the division of the parent-nucleus being homotypic, the chromosomes which go to constitute the nucleus of each daughter-cell are alike both in number and in nature, and exactly repeat the constitution of the parent-nucleus.

In the more complicated cases of propagation by bulbils, cuttings, &c., the development of the new individual, or of the missing parts of the individual (roots, &c.), may be ascribed to the presence in the bulbil or cutting of the necessary pangens.

REPRODUCTION BY GONIDIA

In this case a single cell gives rise to a complete new organism resembling the parent. The inference is that the gonidium is a portion of the parental germplasm, in which all the necessary pangens have been accumulated.

REPRODUCTION BY SPORES

In this case, also, an entire organism is developed from a single cell, but with this peculiarity that the resulting organism is unlike that which bore the spore, a peculiarity which has not yet been explained.

It has been already stated that the development of true spores involves meiosis, and this process is no doubt related to the behaviour of the spore on germination; but the nature of this relation remains obscure. It might be assumed that, as the result of meiosis, the nucleus of the spore receives only gametophytic pangens. But the assumption is rendered impossible by the fact that the spore gives rise to a sexual organism, the reproductive cells of which, after the sexual act, produce a sporophyte.

Clearly sporophytic pangens must be present as well in the spore as in the gametophyte and in its sexual cells. It can only be surmised that they exist there in a latent condition, dominated, as it were, by the gametophytic pangens.

SEXUAL REPRODUCTION

Here, again, as yet unanswered questions present themselves. The essence of a sexual cell is that it cannot give rise by itself to a new organism, it is only truly reproductive after the sexual act: this peculiarity is just what constitutes its sexuality. Minute investigation has not yet detected any essential structural difference between a sexual cell and a spore; on the contrary, the results so far obtained have established that they essentially agree in being post-meiotic (haploid). Why then do they differ so fundamentally in their reproductive capacities?

Again, sexual cells differ in sex; but there are as yet no facts to demonstrate any essential structural difference between male and female cells. What is known about them tends to prove their structural similarity rather than their difference. But it is possible that their difference may be chemical, and so not to be detected by the microscope.

The normal *sexual act* has been described as consisting in the fusion, first, of two cells, then of their nuclei, and finally, often after a long interval, of their chromosomes and of their chromomeres in meiosis. What causes determined these fusions is a question that is only partly answered.

It is known in certain cases (e.g. ferns and mosses) that the male cell is attracted to the female by chemical substances secreted for the purpose by the female organ; that it is a case of *chemio-.taxis*. Probably this is more common than experiment has yet shown it to be. It is quite conceivable that the consequent cell-fusion, as also the subsequent fusions of nuclei and of chromosomes, are likewise cases of chemiotaxis, depending upon chemical differences between the fusing structures.

The sexual process can only take place between cells which are related to each other in a certain degree; that is, it depends upon *sexual affinity*. It is the general rule that it takes place between cells derived from different individuals of the same species; that is, *cross-fertilization* is the rule. This is necessarily the case when the male and female organs are developed upon different individuals, when the plant is said to be *dioecious*. When both kinds of organs are developed upon the same individual (*monoecious*), self-fertilization may and often does occur; but it is commonly hindered by various special arrangements, of which *dichogamy* is the most common; that is, that the male and female organs are not mature at the same time.

But though these arrangements favour cross-fertilization, they do not absolutely prevent selffertilization. In some cases, *cleistogamic* flowers, for instance, self-fertilization alone is possible. The general conclusion is that though cross-fertilization is the more advantageous form of sexual reproduction, still self-fertilization is more advantageous to the species than no fertilization at all.

In considering this subject, it must be borne in mind that the terms used have different meanings when applied to certain heterosporous plants from those which they convey when applied to isosporus plants.

In the latter cases their meaning is direct and simple: in the former it is indirect and somewhat complicated. In heterosporous plants generally the actual sexual organs are never borne upon the same individual, there is always necessarily a male and a female gametophyte; so that, strictly speaking, self-fertilization is impossible. But in the Phanerogams, where there is a process preliminary to fertilization, that of *pollination,* which is unknown in other plants, the terms and the conceptions expressed by them are applied, not to the real sexual organs, but to the spores.

Thus a dioecious Phanerogam is one in which the microspores are developed by one individual, the megaspores by another; and again, self-fertilization is said to occur when the microspores (pollen) fall upon the stigma of the same flower; but this is really only *self-pollination.*

To return to the sexual process itself. Whatever its nature, two sets of results follow upon the sexual act - a zygote is formed, which is capable of developing into a new organism, from two cells, neither of which could so develop; the hereditary sporophytic characters of the two parents are possessed by the organism so developed. These two results will now be considered in some detail.

1. *The Relation between the Sexual Act and Reproductive Capacity*: *In* the early days of the discovery of the sexual process, it was thought that the capacity for development imparted to the female cell was to be attributed to the doubling of its nuclear substance by the fusion with the male cell. Reproductive capacity does not, however, depend upon the bulk of the nuclear substance, for a spore, like an unfertilized female cell, contains but the x number of chromosomes, and yet it can give rise to a new organism.

Again, it has been observed that a non-nucleated fragment of an oosphere of *Cystoseira* can be "fertilized" by a spermatozoid and will then grow and divide to form a small embryo, though it necessarily contains only the x number of chromosomes. From this it would appear that some stimulating influence had been exerted by the male cell, and it is probably in this direction that the desired explanation is to be sought. Some important confirmatory facts have been recorded with regard to certain animals (sea-urchins).

It has been observed that treatment with magnesium chloride will cause the ova to grow and segment; and similar results have been obtained by treating the ova with a watery extract of the male cells. Hence it may be inferred that the male cell carries with it, either in its cytoplasm (kinoplasm), or in its nucleus, extractable substances, perhaps of the nature of enzymes, that stimulate the female cell to growth.

The stimulating effect of fertilization is not necessarily confined to the female cell; very frequently adjacent tissues are stimulated to growth and structural change.

In a Phanerogam, for instance, the whole ovule grows and develops into the seed: the development of endosperm in the embryo-sac is initiated by another nuclear fusion, taking place between the second male nucleus and the endosperm-nucleus: the ovary, too, grows to form the fruit, which may be dry and hard or more or less succulent: the stimulating effect may extend to other parts of the flower; to the perianth, as in the mulberry; to the receptacle, as in the strawberry and the apple: or even beyond the flower to the axis of the inflorescence. Analogous developments in other groups are the calyptra of the Bryophyta, the cystocarps of the Red Algae, the ascocarps of the Ascomycetes, the aecidia of the Uredineae.

2. *The Relation of the Sexual Act to Heredity*: *The* product of the sexual act is essentially a diploid cell, the zygote, which actually is or gives rise to a sporophyte. The sexual heredity of plants consequently presents the peculiar feature that the organism resulting'; from the sexual act is quite unlike its immediate parents, which are both gametophytes. But it is clear that the sporophytic characters must have persisted, though in a latent condition, through the gametophyte, to manifest themselves in the organism developed from the zygote.

The real question at issue is as to the exact means by which these characters are transmitted and combined in the sexual act. There is a considerable amount of evidence that the hereditary characters are associated with the chromomeres, and that it is rather their limn-constituent than their chromatin which is functional: that they constitute, in fact, the material basis of heredity.

From this point of view it is probable that the last phase of the sexual act, the fusion of the chromomeres in meiosis, represents the combination of the two sets of parental characters. What exactly happens in the pseudo-chromosome stage is not known; at any rate this stage offers an opportunity for a complete redistribution of the substance of the chromomeres - in other words, of the parental pangens. It is a striking fact that, in the subsequent nuclear division, the distribution of the chromosomes derived from the male and female parents seems to be a matter of indifference: they are not equally distributed to the two daughter-nuclei.

The explanation would appear to be this, that they are not any longer male and female as they were before meiotic fusion; and that it is because they now contain both male and female nuclear substance that their equal distribution to the daughter-nuclei is unimportant.

The nature of this redistribution of the substance of the chromomeres is still tinder discussion. Some regard it as essentially a chemical process, resulting in the formation of new compounds: others consider it to be rather a physical process, a new material system being formed in the rearrangement of the pangens; here it must be left for the present.

The various ways in which the parental characters manifest themselves in the progeny are fully dealt with in the articles Heredity, Hybridism, Mendelism. It will suffice to say that the progeny, though maintaining generally the characters of the species, do not necessarily exactly resemble either of the parents, nor do they necessarily present exactly intermediate characters: they may vary more or less from the type. It is an interesting fact, the full significance of which has not yet been worked out, that, as a rule, plants that vary profusely are those in which the characteristic 2X number of chromosomes is high.

Brief reference may be made to the cases of abnormal sexual or pseudo-sexual reproduction described above under *Apogamy*. Taking first the cases of true apogamy, there is clearly no need for any sexual process, for, since no meiotic division has taken place, the gametophyte is diploid; its cells, whether vegetative or contained in female organs, possess the capacity for both development and the transmission of the sporophytic characters.

It is not remarkable that such a gametophyte should be able to give rise directly to a sporophyte; but it is remarkable, in the converse case of apospory, that a sporophyte should give rise to a diploid gametophyte rather than to another sporophyte. In the latter case the tendency to the regular development of the alternate form appears to override the influence of the diploid nucleus.

Turning to the various forms of pseudo-apogamy, there are first those in which fusion takes place between two apparently female organs (some Uredineae; Christman), and those in which it takes place between nuclei within the same female organ (*Humaria;* Blackman). If these are to be regarded physiologically as sexual acts, it must be inferred that the fusing organs or nuclei have come to differ from each other to some extent; for it is unthinkable that equivalent female organs or cells should be able to fertilize, or to be fertilized by, one another.

There are finally those cases in which apparently vegetative cells take part in the sexual act, as in *Phragmidium*, where the female organ fuses with an adjacent vegetative cell, and in the fern-prothallium, where the nuclei of two vegetative cells fuse. They would seem to indicate that vegetative cells may, in certain circumstances, contain sufficient germ-plasm to act as sexual organs without being differentiated as such.

An interesting question is that of the origin of apogamy. It is no doubt the outcome of sexual degeneration; but this general statement requires some explanation. In certain cases apogamy seems to be the result of the degeneration of the male organ; as in *Humaria,* where there is no male organ, and in *Lachnea,* where the male organ is rudimentary.

In others, as in the Uredineae, it is apparently the female organ that has degenerated, losing its receptive part, the trichogyne; the male cells (spermatia) are developed normally, and there is no reason to believe that they might not fertilize the female organ were there the means of penetrating it. In yet other cases the degeneration occurs at a different stage in the life-history, in the development of the spores.

In the apogamous ferns investigated, meiosis is suppressed and apogamy results. In the heterosporous plants which have been investigated (e.g. *Marsilia, Eu-Alchemilla*) it has been observed that the microspores are so imperfectly developed as to be incapable of germinating, so that fertilization is impossible; and it is perhaps to this that the occurrence of apogamy is to be attributed. This abnormal development of the spores may be regarded as a variation; and in most cases it occurs in plants that are highly variable and often have a high 2X number of chromosomes.

It will be observed that such physiological explanation as can be given of the phenomena of reproduction is based upon the results of the minute investigation of the changes in nuclear structure associated with them. The explanation is often rather suggested than proved, and some fundamental facts still remain altogether unexplained. But it may be anticipated that a method of research which has already so successfully justified itself will not fail in the future to elucidate what still remains obscure.

Chapter 11

Breeding of Plants

The science of genetics, or heredity, is concerned with questions involved in the transmission, from one generation to the next, of all the attributes which gives each its individuality. In short, genetics seeks to explain why sometimes "like begets like," but at other times "begets the unlike." There are two classes of factors which condition the appearance and behaviour of the individual. On the one hand, there are the factors of the environment.

One plant may be more favourably situated, getting more light, more manure or water, and because of this is bigger, leafier, or more productive. Another, less well favoured, suffering from paucity of supplies, imbalance of nutrients, or disease, is smaller or deformed or discoloured. The two plants, had they been grown under similar circumstances, might have appeared and behaved alike. On the other hand, there are differences due to inheritance.

There are tall varieties and dwarf varieties of peas; wheat varieties with either white grain or red grain; clover with hairy leaves or non-hairy leaves. The production of these characters is not due to factors of the environment, but to factors inside the plant which were transmitted to it from its parents. The characters of the plant, then, as we see them are the result of an interaction between the factors of environment and the internal, hereditable, or genetical factors.

The action of the external factors is' temporary, and produces only *fluctuations* in the course of development; the internal factors are passed on to the next generation unaltered. The genetical factors are quasi-permanent units passing from generation to generation, and conditioning in each, definite *variations* between individuals differently endowed.

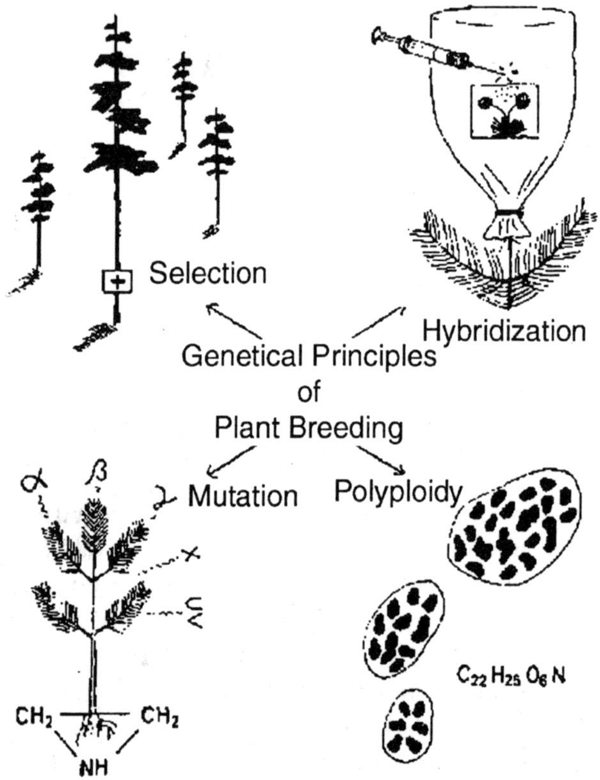

Fig. Plant breeding.

Where genetical questions are involved comparisons between plants should be made only between individuals which have been treated alike, and therefore where variation and not fluctuation has been the cause of any differences between them. Another point which must be clarified is the difference between propagation and reproduction. Propagation is the re-formation of a whole plant from a portion of the body of the "parent."

Only body (somatic) cells are involved; only mitotic or somatic cell divisions occur; there is no fusion of nuclei. Hence, all the cells of the "daughter" plants are constituted or endowed exactly as the cells of the "mother," and contain the same set of factors. No matter whether the propagation is natural, as by runners, stolons, bulbs, corms, etc., or artificially produced by cuttings, grafts, buds, it involves only the splitting up of one body into parts and the establishment of these parts as separate entities.

By propagation, *one individual* is distributed in time and space. An apple variety, "James Grieve," is bred only once. One tree is grown from a true seed. Every unit of "James Grieve" in the world is merely a part of the original tree, or a part of a part of it grafted on to appropriate roots. Similarly, with a commercial variety of potato, the variety is bred once, then tubers (branches of the body) are distributed.

Units produced by propagation are called propagants, or clones. They are identical genetically, and, subject to fluctuation, identical in appearance. Conversely, in reproduction the body cells of the parent do not enter into the process directly; meiotic division comes between the somatic cell and the gamete, which in its chromosomes carries the genetical factors.

The gamete is not constituted like the parental cell, though it derives its constitution from it. Further, in reproduction there is fusion of two gametes at fertilization to give the zygote from which the body of the new generation derives. The two gametes which fuse may have come each from parents of quite different genetical constitution, and so the offspring of the mating may not wholly resemble either parent but each in part. Excluding the question of mutation which will be discussed later, propagation gives complete uniformity in the progeny; reproduction holds out the chance of variation.

CROSSING

The genetical factors which pass from generation to generation must be carried in the gametes, the particular nuclei seen in the pollen grain and embryo sac. To effect a cross between two plants, suipollen must be transferred from the stamens of one to the stigma of the other, precautions being taken that no other pollen is present.

All being well, the pollen germinates and fertilization follows. The two plants so "crossed" constitute the "first parental generation," or Pi, and their immediate progeny, the "first filial generation," or Fi. If the pollen supply used in fertilizing the Fi is restricted to its own spores, and it becomes self-fertilized or "selfed," the next generation is known as Fs.

Simple Cross

By appropriate crossing and selfing the behaviour in heredity of the different factors may be followed. A very simple case is seen in wheat. Some varieties have the spikelets arranged on the rachis, with an interval between. In others, the spikelets are placed quite close together and the "head" is compact.

The one is said to have a "lax head," the other a "dense head." If two plants, one from each of these types, is crossed, the Fi is intermediate as regards this character. When the Fi is selfed, the F2 shows three classes of plant. In one class all the plants have lax heads; in the second they are all dense; while in the third they resemble the Fi, and are intermediate in character.

The cross is made either way, the dense head type or the lax head being used as the female, it matters not; if sufficient plants are grown in Fs, the numbers in the three classes will always bear the same numerical relationship to each other: 25 per cent, lax, 50 per cent, intermediate, 25 per cent, dense; or the ratio 1:2:1.

THE GENE

The mechanism which lies behind this segregation and recombination of factors is simple. It is now established that the genetical "factor" has a material basis in a piece of the structure of a chromosome. The name for this material basis of the theoretical factor is the gene.

Each gene is known to occupy a definite place (the locus) in a definite chromosome. The chromosomes are paired bodies, and in a diploid (somatic) cell-nucleus there are two homologues, each carrying the members of the same gene pairs. The chromosomes of each homologous pair at meiosis part company. This means that every gamete will receive one chromosome from each chromosome pair, and therefore one gene from each gene pair.

At fertilization the chromosome pairs, and therefore the gene pairs, are reconstituted. The gene controlling the arrangement of spikelets on the rachis exists in two forms or potentialities. There is one potential for the lax condition, and the other potential for the dense condition. All the gametes of the lax parent will carry the gene for lax (L), and all the gametes of the dense parent will carry the gene for dense (D).

These two different aspects of the gene pair meet at the fusion of the crossed gametes, and are therefore present in the somatic nuclei of the Fi, where they exert their influence. When the Fi in its turn proceeds to form gametes the chromosome homologues part, and therefore the genes they bear part also. One half of the pollen will receive the gene aspect for lax, the other half for dense. Similarly, 50 per cent, of the ovules will carry "lax," and 50 per cent. "dense."

On self-fertilization of the Fi, it depends entirely on chance whether a pollen grain carrying "lax" meets an ovule carrying "dense" or "lax." Similarly for pollen grains carrying "dense," as there are equal numbers of dense and lax carriers in both the male and female gametes.

The Heterozygote

These hybrid plants whose somatic nuclei develop from a zygote containing both aspects of the gene pair (*e.g.* lax-producing and dense-producing) are known as heterozygotes, and are said to show the heterozygous condition.

On the other hand, plants whose nuclei have two "doses" of the one aspect of the pair are homozygous, and such plants are homozygotes. The homozy-gote breeds true when self-fertilized, for all its gametes are alike. The heterozygote when self-fertilized shows segregation. When it forms pollen and ovules the two aspects of the gene pair part company and, depending how they meet for fertilization, produce plants of different character. The essential point here is that gene pairs segregate at reduction division because their chromosomes segregate.

They recom-bine at fertilization because at that point in the life-cycle the chromosome pairs are reconstituted. This explains why there are three classes in F2, and the fact that when large numbers are grown in this second filial generation the numbers in each class occur with the bounds of reassortment on the basis of pure chance in the ratio of 1: 2: 1.

DOMINANCE

When other characters are considered in a hybridizing experiment, the Fi may not be intermediate but resemble one of the parents. There are bearded wheats and non-bearded. That is, the "chaff" which envelopes the grain may have a long tapering awn (bearded), or this may be absent (beardless). When a bearded wheat is crossed with a beardless one, the Fi is bearded.

The gene for bearded dominates in the Fi. It is said to be dominant to the non-bearded gene, which conversely is said to be recessive. In Fs only two classes appear—beardless and bearded. The numbers of plants in ttyese classes are in the ratio of 3: 1. In Plate 120 specimens from the actual plants in such a cross have been mounted in such a way as to illustrate the behaviour seen in a mating of this type. In the lowest line of plants (the Fa) it will be seen that there are twelve bearded plants to four without beards; that is, the numbers in the two classes are in the ratio of 3: 1.

This is quite simply interpreted when it is remembered that one "dose" of bearded "swamps" or masks the presence of one dose of beardless. It is clear that the 3: 1 ratio is merely the 1:2:1 with the first two classes amalgamated by the effect of dominance. Only one plant (the homozygote) in every three of the bearded class will breed true.

In this particular case the heterozygote produces much smaller awns than those borne by the homozygote. This partial production of the character is called partial dominance. When the heterozygote is quite similar to the homozygote the dominance is said to be total.

Phenotype and Genotype

A plant described on the basis of its external appearance is called the phenotype. If described by reference to the genes inside it is called the genotype.

In the Fs, then, there are bearded phenotypes, some of which (the heterozy-gotes) are genotypically bearded and beardless, while the others (homozygotes) are phenotypically and genotypically beardless.

Genetical Purity

In order to distinguish clearly between a homozygous and a heterozygous bearded, or, in other words, to distinguish between the two possible genotypes, the plants must be self-fertilized and bred-on. The progeny of the two genotypes will differ. On self-fertilization the homozygote produces only one class of plant, or, in other words, it "breeds true."

On the other hand, the heterozygote segregates. This "proving" of the plant by reference to the character of its offspring is called the progeny test. Where only a small number of plants can be grown, it may so chance that following on self-fertilization the recessive form does not happen to be produced. If, however, the suspected heterozygote, instead of being self-fertilized, is crossed with the pure recessive parental type, the chances of the recessive condition appearing in the progeny is increased.

Every pollen grain from a beardless plant carries the gene beardless. This, supplied to a heterozygous plant, ensures that every ovule carrying the recessive gene will produce a beardless individual. In short, the cross of a heterozygote to the recessive ensures a 50/50 distribution in the two classes.

Crossing back to the recessive parental type is a special case of progeny testing called the *back-cross*. The phenomenon of segregation and recombination of these genetical factors was discovered in the garden pea by Mendel. The ratios which he demonstrated are called Mendelian ratios, and the study of their various forms, Mendelism.

Very many other but similar cases have been described in all sorts of organisms by many investigators. Sometimes dominance appears, sometimes the Fi is intermediate. When the parents of a cross differ in two pairs of characters, both showing dominance, four classes of plant appear. This is due to two 3:1ratios being superimposed. Re-examination of the photograph will show that the left-hand awnless parent (variety Iron) has also white chaff while the right-hand bearded parent (variety April) has red chaff.

These two parents differ in two characters—awnedness and chaff colour. The Fi is seen to resemble the " April" parent in that it is both bearded and red-coloured in the chaff. When the Fi proceeds to form gametes the two pairs of genes involved segregate quite independently of each other, and therefore produce equal quantities of pollen grains in four different classes, each class differing from the others in gene content. Similarly there are four classes of female gamete.

Chromosome Gene

A plant shows hundreds of units of character and must therefore have hundreds of pairs of genes, yet it has only a limited number of chromosomes in each nucleus; in bread wheat there are twenty-one pairs in the soma or twenty-one individuals in the gamete.

Many other plants have fewer chromosomes, yet these bodies have to carry all the genes. It is known that each chromosome carries a number of genes, and these are distributed up and down the chromosome in linear order like beads on a string. Each chromosome always carries the same gene set. The distance between neighbour genes on a chromosome is constant, and the order in which they occur from one end to the other never varies in normal cases.

Linkage would prohibit the possibility of new combinations being bred, were it not for the fact that at syndesis in meiosis, the chromosomes of homologous pairs become closely intertwined, and in separating give the appearance of having broken and united with an exchange of parts between the two members of a chromosome pair.

The points where they cross-over each other in the intertwining, called a *chiasma,* can be seen under the microscope. On the left is the original chromosome pair; next the two intertwined; next the two chromosomes are shown reformed with exchange of parts, while on the extreme right are the chromosomes as they go forward to complete mieosis and pass to the gametes.

MUTATION

It may be asked now how a plant comes to have two aspects of a character or how a character pair comes to be. It is known that the wild-type gene as it occurred originally may alter its nature. The alteration produces a new expression of the character. The process, or act of alteration, is called mutation. The new type of plant so produced is called a mutant or a mutation. A gene in the majority of cases mutates only once, and thus gives a character pair, the wild type and the mutant. The same gene, however, may mutate again and give three forms of the gene—the original wild type and two mutants.

This provides three expressions of the character. Only two of these three can be present in any diploid somatic nucleus, and only one in any gamete. A case may be taken from the haricot bean, plants of which may have green foliage and green pods (Gr) or green foliage and yellow pods (G), or yellow foliage and yellow pods (g).

When Gr is crossed with G, the Fi shows green pods and foliage. The Gr condition is dominant to the G condition. When the G type is crossed with the g, the green foliage and yellow pods condition is dominant. Thus we see a series of decreasing dominance—Gr dominant to G, and either of these dominant to g. Such a graded set is called a multiple-allelomorphic series.

All sorts of quantitative and physiological conditions are known where the genes for the characters belong to such a series. In some series there may be more than three aspects, for cases are known where the gene has mutated five or six times, and six or seven character expressions exist.

HETEROPLOID

Another aspect of cytology affecting Mendelian ratios must now be referred to. It sometimes happens that at mitotic or meiotic division the daughter nuclei do not separate cleanly in the usual manner. In these cases the chromosomes, either as a whole set or as part of a set or even individual members, lag behind.

The laggards which should have gone to one daughter nucleus are included in the other. When this occurs, nuclei are formed with a number of chromosomes different from that normally found in the species. A nucleus with such an aberrant number of chromosomes or a plant in whose nuclei they occur is said to be heteroploid. If a whole chromosome set is replicated, the heteroploid is said to be balanced or *euploid.*

If only a part of the set is replicated, the nucleus is of the unbalanced or *aneuploid* class. Most attention is directed to those cases where whole sets are concerned, that is, to the euploid form. Where one replication of the normal diploid has taken place, and the homologues are in threes, the plant is said to be *triploid.*

Where the diploid has doubled and the homologues are in fours, the plant is said to be *tetraploid.* Plants are known where the body cells contain only half the normal diploid number; that is, the number normal to the gamete of the species. These plants are known as *haploids.* There is thus an ascending series in the terminology: haploid, diploid, triploid, tetraploid, pentaploid, hexaploid, and upwards.

These have been formed in nature as the result of some unknown stimulus, or artificially in cultivation by the use of such drugs as colchicine.

HETEROPLOID'S EFFECT

Giantism

Replication of chromosome sets often results in the production of a bigger (giant) form of plant. This in itself may be very useful as, when it occurs in a plant of a forage crop.

Sterility and Fertility

Heteroploidy may be very useful in regard to sterile hybrids. These hybrids often result from crosses between plants which are not very closely related, but close enough to form a new hybrid plant. The chromosomes derived from the two parents are sufficiently homologous to form a mixed nucleus in Fi, but not so as to behave normally at meiosis. The hybrid fails to form viable gametes. Such hybrids or "mules" are sterile, and therefore, unless they can be propagated successfully and economically, are of little or no use in practice.

Replication of the mixed sets of chromosomes, if it occurs in such a sterile hybrid, may render it fertile.One of the clearest examples of this is furnished by a non-agricultural plant. *Primula verticillatay* with nine pairs of chromosomes in the somatic nuclei, can be crossed with *Primula floribunda,* which also has nine pairs of chromosomes. The gametes of each, therefore, contain nine chromosomes, and these form the hybrid. None of the nine floribunda chromosomes are homologous with the nine verticillata members, and therefore successful meiosis in the hybrid is barred.

Viable gametes are not possible, and the plant is sterile. The sterile hybrid formed a euploid and immediately became fertile. Schematically, the mating, the chromosomes being symbolized by the initial letter of the species from which they are derived.By replication, each V chromosome obtains a homologue and similarly each F chromosome.

Normal meiosis is possible and gamete production follows. Many species not closely related can form hybrids which are sterile. It is probable that many of these will be rendered fertile when duplication occurs. Wide crosses as between, a wild grass like couch and wheat, or between turnip and cabbage, are important because the new gene complex brought into being results in more than a mere addition of the characters of one parent to those of the other.

Genes even in a normal diploid soma interact on each other to some extent, and when two complete gene complexes interact the effect is greatly increased, so that a new plant different from both parents is brought into being. This is paralleled in chemistry when, the union of sodium, a white silvery metal, with chlorine, a green gas, produces a new substance, sodium chloride with characters quite different from the parent substances. The formation of heteroploids has occurred in nature. Our modern bread wheat as it was evolved is a complex hexaploid of diverse parentage with three sets of seven, that is, twenty-one, chromosomes in the gamete, and twenty-one pairs (forty-two individuals) in the somatic nucleus.

One interesting result of heteroploidy occurs when the component sets of chromosomes each contribute to the new plant similar genes. A case in point arises in connection with the production of the red colour in the grain of bread wheat.

Here, if an appropriate variety with red grain is crossed with a variety having white grain, a 3: 1 ratio red. If a different red-grained variety had been used as a parent, a 15: 1 ratio red: white in Fs would have been obtained. If yet another variety had been employed, the ratio of red: white in F2 would have been 64: 1.

Results such as these indicate that three pairs of genes are involved in the production of grain colour. Furthermore, each pair must be located on different chromosome pairs, because the genes segregate and recombine freely. In short, there must be six chromosomes (three pairs) in each hexaploid somatic nucleus, all of them carrying a grain-colour gene. Cytological evidence shows the bread wheat to be hexaploid.

Multiple genes (polymeric series) of this type become important when they affect quantitative characters, and each dose of, the dominant produces an increase in the character. Sugar content of sugar beet depends on a set of replicated genes, and every additional "dose" of gene produces an increase of sugar content.

SELECTION OF PLANT

Cytological aberration and genetical hybridity provide plant populations variable in character. Variation in a crop, no matter how induced, provides the plant selector with his material. *Two* inbred strains of Maize on eitiier side of the vigourous hybrid got from their mating.

A new combination of genes or gene complexes can be the starting point of a new variety waiting to be selected and multiplied. There are three methods of selection.

Pure Lines

The first consists of selecting single plants, self-fertilizing them, and maintaining the progeny of each as a separate "line." In effect this subjects each selection to the progeny test. The best line which breeds true can be retained and the others scrapped. This method is known as single-plant selection. If the original parent was homozygous for the characters observed, a non-varying or *pure line* will have been formed. A pure line consists of the descendants of one homozygous individual, self-fertilized and never cross-fertilized. This amounts to *genotypic* selection.

A line pure for *all* genes is probably not possible, but a line pure in all the genes that matter is obtainable in practice. It is to be noted that, so far as concerns the genes for which the line *is* pure, further selection can have no effect. A pure line, then, by definition is non-varying. Apart from environmental fluctuation and the possibility of mutation, no alteration in appearance or behaviour is possible.The second method of selection is to choose all the plants in a crop which appear to conform to a predetermined standard.

Such is mass or *phenotypic* selection. This is often useful with plants which are individually self sterile, or where the character is quantitative and depends on a comparatively large series of polymeric genes. Mass selection gives an immediate effect with a large bulk in the first generation. As generation follows generation, if re-selection is not practised on each, the effect wears off, and the strain "runs back" to an average.

Mass selection is often possible on large-scale lines. In many countries growing crops are inspected officially, and where a crop conforms to a definite standard the grower receives a certificate for it. When such a certification scheme is widely patronized it amounts to a large-scale mass selection. The third method of selection is not to select out the good plants, but to pick out from a crop the wrong ones or "rogues." These are the plants which do not conform to standard.

Roguing has a potent influence when the rogue is due to a dominant gene, and when the desired character is due to a recessive. Roguing out a recessive is not really effective in practice, for it leaves in the crop the heterozygotes. In this latter case it usually pays better to develop a pure line following a progeny test.

STERILITY

Self-fertilization and cross-fertilization have been referred to. These are often barred by sterility. Almost any cabbage plant will illustrate the point. If its own pollen only is allowed access to its stigmas a cabbage will set few or no seeds. If pollen from other cabbage plants gets access to the stigma, fertility is complete.

This proves that its ovules are fully functional. Its pollen is fully effective in producing seed on most others of its fellows. The gametes of the plant accordingly are fully viable but self-incompatible. It is now known that the tube of a pollen grain carrying any Sallelomorph can hardly pass through a style carrying this same allelomorph in the pair resident in the cells. A similar mechanism affects many varieties of fruit tree. If an orchard has been planted only with such a self-incompatible variety, every one of the trees produced by vegetative propagation will have exactly the same gene complex.

Hence, every stigma will contain incompatible genes in common with the pollen supply and fertility will be barred. Such an orchard must be interplanted with a few trees of another variety which is known to be capable of supplying suipollen. Many plants are known to have a cytological constitution which does not permit of normal meiosis. The plants are usually of hybrid origin. Many fruit trees are of this nature and do not form viable gametes.

VIGOUR LOSS

Where sterility is no bar, and continued self-fertilization is practised, another disability may show itself. The vigour of the progeny derived through continued self-pollination falls progressively generation by generation.

In maize, after a few generations of self-fertilization the inbred lines become weak, and in extreme cases may fail completely and die out.

Hybrid Vigour

Curiously, when two inbred but unrelated lines are crossed the F1 is often extremely vigourous. Out-breeding brings back vigour. This vigour in a hybrid has been called *heterosis* or *hybrid-vigour*. It shows only in the F1 and does not appear in F2 when the F1 is self-fertilized.

When appropriate inbred lines are used as parents, the Fi is often very much more vigourous and yields much more than any commercial lot. Heterozic or Heterozygous maize seed is now sold for crop production. In fact, for this crop, practically all progressive growers sow only hybrid vigourous seed.

All of the Fi crop goes for consumption as food, etc.; none is retained for sowing. Each year the Fi is made anew from inbred parental lines which are known to give the maximum effect when crossed. The commercial production of heterozic maize seed is simple. The two parental lines are interplantcd in rows.

At flowering time all the male inflorescences are removed from the plants designed as the seed parent. It is therefore pollinated only by its neighbour, and all the "seed" produced by it is of hybrid origin. In any crop produced by vegetative propagation the vigourous hybrid need be produced only once. All clones arc genetically identical and just parts of the Fi itself.

They retain all its characteristics in full effect. This is well seen in the poplar tree used for wood pulp. Two parents arc crossed. If the progeny is highly vigourous, suckering is induced. The suckers are planted out. The trees from such suckers grow very fast and harvest is brought nearer by a number of years.

Chapter 12

Diseases of Plants

Plants and potential parasites often exist side by side without any appreciable amount of disease resulting. A slight change in the environment, rendering it either less favourable to development of the host or more favourable to that of the parasite, may then lead to an extensive outbreak of disease. In such an outbreak the active agent of disease is the parasite but the immediate cause of its activity is the change in the environment.

Hence it is often necessary to decide whether control measures are likely to be most effective in a direct attack on the parasite, as by the use of fungicides, or indirectly by modifying some environmental factor, such as temperature, humidity or nutrition of the host. Accurate diagnosis is in all instances an essential prerequisite in a programme of control. It may have to be decided whether a valuable consignment of plants has been damaged by parasitic fungi or by heating in transit.

Moreover different fungi have very different life histories, and control measures are only likely to be effective if they are directed against some particularly vulnerable period in the life of the parasite. Among the causes of plant disease the first to be considered is the hereditary factor or gene, which is a part of the plant itself and may therefore be regarded as an internal source of disease.

Next come the environmental factors responsible for non-parasitic diseases, especially in cultivated plants grown under unnatural conditions, with excessive, deficient or unbalanced supplies of nutrients.

Finally there are the many living organisms that cause parasitic infections or contageous diseases.

DISORDERS OF INHERITANCE

Researches in heredity have brought to light numerous instances where a change in the inherited constitution of the plant has led to the appearance of variegated leaves, deformities—sometimes of commercial value, as in the fasciated form of cockscomb *(Celosiacristata)*—or even complete absence of chlorophyll. In the latter case the seedling inevitably dies as soon as the food stored in the seed has been exhausted.

Change in the hereditary constitution may also take place in individual buds (bud sports). Something of this kind is probably responsible for the appearance of the " wilding " type of growth in potatoes. "Degeneration " or " running out " is a popular expression used to cover any incomprehensible decline in vigour of a crop. Originally the terms were applied to the unfavourable and weakening effect that continual vegetative propagation was imagined to have on certain kinds of plants.

Subsequently it was thought that continued cultivation under unsuitable conditions could have a cumulative injurious effect on the constitution of a plant. Critical investigations have yielded little or no evidence in favour of either of these hypotheses. Degeneration in vegetatively propagated crops is due in most instances to the spread of virus diseases, especially in potatoes, strawberries and raspberries.

These diseases are seldom transmitted by the true seed but their casual agents permeate the entire vegetative body of infected plants and are inevitably transmitted by all their tubers, runners, cuttings or offsets. Other so-called types of degeneration are expressions of self-sterility or the result of inbreeding.

FACTORS DEPEND ON ENVIRONMENT

WATER SHORTAGE

Transpiration or evapouration of water from the leaves and other above-ground parts is an essential function in land plants. Not only is it one of the factors which brings about the ascent of sap but, owing to the large consumption of latent heat in changing liquid water into water vapour, transpiration helps to cool the plant. In hot sunny weather there is intense radiation from the surface of bare soil which scorches the moisture still more out of herbaceous plants.

The effect varies greatly between different plants. Potato leaves wilt helplessly, but beetroots, owing to their great ability for obtaining during the night the moisture they require from the soil and dew, have their leaves as erect as ever next morning. Sunflowers transpire especially through their lower leaves which wilt first; begonias, on the other hand, mainly through their flowers. There is little restriction to evapouration from the surface of most petals and it is a common experience that large showy flowers like chrysanthemums and roses wither very readily when cut.

In extreme drought, plants, and even deciduous trees, shed their leaves. A second crop of leaves is then produced when the drought is over. Similarly in very dry summers apple trees shed a large proportion of their young green apples. The dry air of living rooms tends to production of " long-legged " pot plants with only a few leaves near the top of their stems. Out of doors a dry wind may shrivel the edges of leaves, especially of the newly opened leaves of pear, apple and red currant in the spring. Such marginal scorching becomes more pronounced the further one gets from shelter. Marginal scorch of fruit-tree leaves is often, however, a sign of potassium deficiency, especially later in the season.

Less often, drought injury becomes apparent in the centre of leaves, as in tomatoes, where it appears as angular or irregular leaf spots. Normal autumnal leaf fall of deciduous trees is less a protection against cold than against drought.

Leaves of evergreens have anatomical devices to restrict and control transpiration but even in conifers dry spring winds, coming before the roots have begun to function adequately, too often give rise to the well-known " red fir trees."

The terminal buds generally survive on the branches of such trees and later produce the normal shoots with young green needles of the current year, behind which is a bare zone where the needles of one or two previous years have been scorched, dried and fallen off. Premature shedding of flower buds or flowers, especially common in begonias, epi-phyllums and camellias, is often a sign of drought.

In the lily and amaryllis families the flowers react to drought in a peculiar way, by adhesion of the tips of the petals.Fruits also have their own characteristic ways of reacting to a shortage of water. Tomatoes develop blossom-end rot, pears stone cells in the flesh, and apples bitter pit. During severe drought potatoes develop tubers as soft as indiarubber, either at the heel end only or throughout, with a tendency to a sprout in early autumn soon after lifting.

Indirect effects of drought are the encouragement it gives to some insect pests, especially aphides, and to a few parasitic fungi, notably the powdery mildews. Absorption of certain nutrients from the soil is also hindered, especially of minor elements like boron. Such deficiency diseases as heart rot of sugarbeet and canker or black spot of garden beet are thus often associated with dry weather. The boron-deficiency diseases of apples are similarly encouraged and were hence known as " drought spot" before their primary cause was understood.

Because of the severe losses so often caused by drought various methods of artificial water supply have been devised, but there are other less direct ways of mitigating the effects of dry weather. These include destruction of weeds, keeping the surface soil loose by hoeing, planting shade crops or supplying artificial shade, mulching the soil, etc. When plants have been raised under glass it is important to harden them off gradually if they are later to be exposed to a dry atmosphere.

Water in Excess Amount

Roots must be able to breathe, and in waterlogged soil. the fine absorbing roots of most plants are choked and die so that absorption of nutrients and even of water is impeded. Water plants have a system of internal air channels which supply the roots with oxygen from above the water level.

Potato tubers are very sensitive to waterlogging and may be killed after a day's complete immersion in water. Occasionally tubers planted in wet soil do not rot but become glassy, very hard, and fail to sprout. Detrimental chemical changes may take place in saturated soil and the population of soil microorganisms may also be adversely affected. Changes of this kind can only be slowly corrected after thorough drainage. Besides producing the suffocated, brown, sour-smelling roots found in waterlogged soil, excessive water supply, whether in a plantation or a flower pot, is often manifested by yellowing of the leaves.

This indicates, amongst other things, that the nitrogen in the soil is not being absorbed by the suffocated roots and partly that it has been washed or leached out of the soil and lost. In house plants this " drowning " sometimes induces a bright yellow colour in the foliage (aurigo), especially in dracaenas and the like. An instance of high death-rate among fruit trees and bushes in Kent is described on Often, especially in greenhouses, the soil contains only a moderate amount of water but, owing to inadequate ventilation, the air becomes overcharged with moisture.

If this condition persists the check to normal transpiration leads to a number of peculiar outgrowths on the above-ground parts of the plants. Oedematous eruptions composed of greatly enlarged cells full of sap appear on vines and tomatoes. In potatoes grown on benches under glass this swelling and eruption of the leaf cells may occur to such an extent that the whole leaf tissue turns mealy and falls away in patches leaving large holes, while the whole leaf becomes curled and deformed.

Such a condition appears very alarming but with sunshine and proper ventilation fresh normal leaves are quickly produced and the plants recover. On leaves of *Pelargonium zonale* wart-like intumescences may be produced by enlargement of the cells which ultimately collapse and die leaving a scab of corky tissue.

The inner surface of pea and bean pods is often covered by a layer of such inflated cells, looking like a coating of white mould. Cacti may react to excessive humidity by becoming in places clear and glassy. In damp situations the breathing pores or lenticels in the bark of many trees become enlarged into corky warts. The same kind of lenticel proliferation may happen in potato tubers produced in wet, though not completely saturated, soil and is often mistaken for a parasitic infection.

In some trees the loosening of the texture of the bark may break out in blisters or involve large patches of the trunk or branches which become covered with a loose corky powder the colour of tan, called by the Germans " tan disease" *(Lohkrankheit)*. The condition is probably harmless and may disappear when the trees are transplanted to drier situations but it is disfiguring and hence reduces the market value of nursery stock.

It is advisable to be very careful in winter-spraying trees which show these symptoms. When several big branches are removed together from an old tree the pressure of the rising sap within the trunk may become so high that cracks appear in the bark. It is therefore best to remove undesirable branches a few at a time. An excess of water in the soil favours attack by many parasites, especially where a motile swarm-spore stage occurs in their life history.

Hence powdery scab in potatoes occurs mainly in ill-drained soils and is only prevalent in the wetter parts of the country and on heavy soils; club root of crucifers is also encouraged by a waterlogged soil or a heavy spring rainfall.

Some bacterial diseases such as potato black-leg are also favoured by a high soil-moisture content; black-leg is usually only prevalent in ill-drained parts of a field. High atmospheric humidity, followed by condensation of moisture on the roof and dripping on to the plants, frequently leads to an outbreak of grey mould *(Botrytis)* diseases in glasshouses and conservatories.

Even out of doors the still damp air of warm muggy days favours an outbreak of many diseases, especially if it is accompanied by dew at night. Well-known diseases encouraged in this way are chocolate spot of beans and potato blight. Sudden changes in water supply from drought to great humidity damage many garden products.

Splitting in plums and other tree fruit is all too common after heavy rain so also are cracked gooseberries, fallen hazel nuts, split tomatoes, telescoped apples, cracked heads of cabbage and carrots and parsnips split right to the core. The potato tuber seems particularly unfitted to cope with a sudden increase in the water supply. In mild cases the skin splits all over in a fine network of cracks, but often the splitting goes much deeper and opens the way to various tuber rots. Splitting generally takes place before lifting but it may also occur during or after lifting.

Sometimes a clear transparent patch of tissue develops at the centre of the tuber which soon dries and cracks to form a lens-shaped hollow, lined with dead brown tissue (hollow heart). This condition is most prevalent in big tubers and is often associated with heavy nitrogenous manuring and wide spacing of the plants. Close planting, balanced manuring and thorough and early cultivation of the potato bed should prevent damage of this kind.

Boron deficiency or infection with some viruses may also increase the tendency to splitting. When heavy rain follows drought towards the end of the growing season potatoes are very liable to second growth. The already-formed tubers develop swollen ends or fantastically shaped outgrowths and the stolons produce an abundance of small new tubers, often several in a row; this last condition is also, however, often associated with the " wilding " condition irrespective of changes in water supply.

The condition known as Jelly End Rot of potato tubers fc apparently due to the roots continuing to function and supplying water to the tubers after the haulm has died or been killed. Such tubers grow out at the rose end and withdraw starch from the tissue at the heel end which becomes glossy and ultimately decays. Under glass a sudden change from drought to high humidity leads to oedema, bitterness in cucumbers, breaking of chrysanthemum stalks just below the head and splitting of the calyx in carnations.

FACTORS OF CLIMATE

Damage Caused by Hail

Hail damage, varying in extent with the size of the hailstones and the intensity of the storm, is to be regarded as a purely mechanical injury. The local cooling of the tissues due to temporary contact with the frozen drops is of little or no consequence. When the hailstorm occurs during the blossoming of fruit trees, the blossom is usually destroyed and the loss irreparable.

Once apple fruit has set and begun to grow the damage depends on the force of the blow. Light hail that does not break the skin will cause only small sunken spots below which dry, dead tissue will be found. If the skin is broken *Monilia* often gains a footing or large scars are formed from which pieces of dead tissue gradually fall away or, in the worst cases, the whole apple may be knocked to pieces by the hail. Hail causes dead dry spots on gooseberries and strawberries if it falls when the fruit is half ripe, but here also ripe fruit is liable to be smashed to pieces in a hailstorm.

Large leaves may be badly torn by hail and sometimes both leaves and fruit are shed. Plants especially liable to hail damage are: rhubarb, beet, leek, onion, cucumber, spinach and lettuce. Rows of peas may be beaten flat and appear to be ruined by a hailstorm, but with good growing weather they recover with amazing rapidity though they may not altogether lose the marks of hail on the pods.

Badly battered potatoes may similarly recover and yield quite well in the end. The bark of trees may be marked by small, well-defined hail scars as a rule the fresh wounds are deep enough to expose the bast or wood, but in the course of a few years the wounds become beautifully sealed by ingrowth of callus tissue from their edges. This often happens in apple trees, but in trees with scaly bark, like the cherry, the wounds are larger, and soft bark, like that of the elder, may be peeled off in large flakes. Crop must be expected from plants whose leaves have suffered extensive injury from hail.

Heavy hailstorms may also beat the surface layers of the soil together and necessitate additional cultivation or hoeing. The one-sided character of hail lesions, due to the hailstones having been driven from a particular direction, is a help in diagnosing hail damage and in differentiating it from somewhat similar injuries due to other causes; especially spots and shot-holes caused by fungi and bark scars due to chafing.

The amount of loss from any particular storm cannot be crop must be expected from plants whose leaves have suffered extensive injury from hail. Heavy hailstorms may also beat the surface layers of the soil together and necessitate additional cultivation or hoeing. The one-sided character of hail lesions, due to the hailstones having been driven from a particular direction, is a help in diagnosing hail damage and in differentiating it from somewhat similar injuries due to other causes; especially spots and shot-holes caused by fungi and bark scars due to chafing. Hailstorms form a special hazard in seed production, especially towards ripening, partly estimated immedi-ately, because they beat down the plants and partly parisons with the yield of undamaged crops because of the great shedding of seed that have been made and when it has been seen may result.

EFFECT OF TEMPERATURES

Some plants are readily killed by low temperatures. Certain tropical plants even die at 41° F., others not until the air temperature falls far below the freezing point of pure water, 32° F.; but frost, even when it involves formation of ice within the plants, is often less fatal than is generally supposed.

Much so-called " frost damage" is the result of drying out of the tissues; the dry winter winds deprive pine needles and evergreen leaves of so much water that their tissue dies at the tip, round the edges, in patches or even over the whole leaf. If the ground around the roots is frozen they cannot be supplying water to the above-ground parts of the plant or tree. The benefit derived from planting shelter belts often demonstrates the great effect drying winds can have during a frost.

The natural autumnal leaf fall of deciduous trees is to a large extent a protection against water loss in winter. Bud scales protect the tender next year's shoot against drying more than against cold. The final function of the deciduous leaf stalk is to lay down a layer of cork to seal the scar left when the leaf falls; if they are killed by a very early frost the leaves may hang on the trees until far into the winter.

When cold weather comes many plants convert starch, an insoluble food store, into sugars or fats, especially in bark and evergreen leaves. Because the presence of dissolved substances depresses the freezing point of a liquid this may be regarded to some extent as a protection against ice formation in the living cells.

This is not the main reason for storage of reserve foods in the form of sugar, however, as is shown by the facts that sugar beet is not particularly hardy while sugar cane is a tropical plant. In apple trees, raspberries and gooseberries, hardiness seems to be correlated with pentosan content. Ripening or maturing is a comprehensive term covering the changes that take place in autumn in the current year's growth of a tree or bush.

Normal ripening is of great importance in avoiding winter injury. Ripening includes the production of protective substances, a considerable fall in the water content of the tissues and changes in the protoplasm of the living cells. Plant cells do not contain pure water but strong solutions whose freezing point is a good deal lower than 32° F. Ripening is affected by many conditions. It is retarded by an excessive supply of nitrogen, and in raspberry plantations it has been shown that excessive liming weakens the cane's resistance to frost.

Common spruce has been found to suffer frost damage when growing in a very acid soil of pH 3.9, but not in a less acid soil of pH about 5. In New York State some correlation has been found between a shortage of lime and frost injury to apple shoots. Conditions of terrain, such as deep or shallow soil, light or heavy soil, may also influence the amount of frost damage. In fruit farming the earlier ripening of fruit in grass orchards helps it to escape frost. Parasitic diseases may also have an influence on the frost resistance of plants. Hauch and Ravn have shown that oak mildew retards ripening of the shoots and hence increases their liability to injury by frost.

Apple scab reduces the production of starch by the leaves, and so retards ripening of the shoots, and makes frost damage, often followed by canker, more likely. Bolting of celery, beet and other biennial vegetables is in part governed by hereditary characters but the tendency is much increased by spring frosts. Cold air is heavier than warm and it tends to flow towards low ground. In winter if the sky is clear and the air still the cold air gradually collects in hollows and valleys.

In such circumstances it may freeze in the hollows while on higher ground there is no frost. " Frost pockets " of this kind are a well-known problem in fruit growing. The tissues to surfer most from spring frosts are the tender young growing organs full of sap, such, as young potato shoots, young beans and the flower buds of fruit trees. In the latter case, however, blossom blight due to *Sclerotinia* is very commonly mistaken for frost damage.

Flowers of black currants and other fruit buds are liable to be destroyed by frost, while in strawberry blossoms the pistils are blackened so that the fruits do not set. Streaks, rings or spots of corky skin on apple and pear fruit and green blotches round the nose of ripe apples may also be due to frost; so are blistered leaves of apple and other trees and small blisters on young bark. Half-open chestnut and elm leaves may have their exposed surfaces scorched by frost and whole patches of tissue may be killed so that when the leaves finish unfolding they split and fray like feathers. In other leaves only the margins or tips may be scorched, depending on how the leaves are folded in the bud and which parts were exposed when the frost occurred.

Frost cracks may be produced not only in green, juicy tissue but also in wood. A sudden great fall in temperature may lead to such high tension in a tree trunk that it splits from crown to root, with a loud report. When the temperature rises again the crack closes. Often the adjacent cambium and bark are almost undamaged and the wound is healed by formation of a frost callus.

The cracked place may, however, remain weak so that the annual rings covering it are split again in the same place when the next hard winter comes. Great differences in susceptibility to frost damage may exist even between varieties of the same plant, and this is an important factor for the consideration of the plant breeder, in winter broccoli. Among apples Worcester Pearmain, Allington Pippin, Grenadier and Lane's Prince Albert surfer less from frost than Bramley's Seedling and Cox's

Orange Pippin. Conference is notoriously frost resistant among pears and Czar among plums.

Ice formation in the tissues takes place most easily when water has seeped from the cells into the air spaces between them. Freezing of this intercellular water then often leads to the formation of columns of ice, often so small that they can only be seen under a microscope, but they are a useful indication of frost damage.

Some plants do not surfer from ice formation in their tissues, though only a few are like *Larix sibirica* which can stand 30-50° C. of frost for six months of the year. A number of tissues are not damaged by freezing if they are allowed to thaw slowly. Thus, edible roots frozen into the ground should be left undisturbed until the ground has thawed. This may be partly connected with the phenomenon of super-cooling, i.e. there may be no ice formation even when the temperature has fallen several degrees below the freezing point of the sap, but if the plant is then shaken the whole tissue freezes suddenly.

Of course, if the tissues have not merely been frozen stiff but also frozen to death no caution in thawing is of any use. Death of the living cells by freezing may be due to their being crushed and torn by the expansion of ice when the water in the intercellular spaces freezes. More often it is due to excessive withdrawal of water from the cells into these spaces as the ice forms.

This withdrawal of water may proceed so far that the concentration of soluble substances in the cell sap reaches an injurious level. Moreover the protein substances of the living cells occur in a colloidal condition, and excessive withdrawal of water may lead to a coagulation of these colloids which is irreversible and cannot be overcome when the ice melts and the cells reabsorb water. Oxidation of some of the constituents of these damaged cells turns them brown or black. When the cells have not been killed they are able to reabsorb the water from the spaces between them provided thawing proceeds slowly.

If it occurs suddenly these spaces become flooded with water, however, and the vital processes of the living cells are impeded by lack of air. Freezing to death may occur either in a whole plant or only in parts of it. Green and juicy parts then become glassy and when they thaw they are limp and the water may be squeezed out of them. All resistance to bacterial and fungus infection is lost. Several weak parasites like *Botrytis, Valsa* and *Nectria cinnabarina* gain a footing in frost-damaged shoots and are able to spread thence into the adjacent living tissues.

Gummosis and die-back of branches may sometimes be the direct result of frost injury, sometimes of frost damage aggravated by feeble parasites of this kind. Sometimes frost-damaged trees burst into leaf but weeks or even years later the affected branches wither or the whole tree may die. Avoiding injury from cold and frost. Mulches of leaves or manure, with branches of evergreens or mats to cover espalier trees, are commonly used to counteract too great variations in temperature on warm days in early spring. Planting shelter belts may be effective in the long run. In places where night frosts are a problem it may be possible to deflect the cold currents by suitably planned hedges (air drainage).

Earthing up in various ways may be undertaken in autumn to give shelter. Fruit blossom may be saved from frost at the critical period by a judicious use of orchard heaters at the rate of 50 to 60 per acre. If this is attempted, however, it is necessary to heat a considerable area. It is useless to attempt to control frost in one acre of a large frost pocket as there will be too much cold air accumulating all round it.

The smoke drifting from a heated orchard may also afford some protection from frost to field crops in the vicinity. In countries with a continental climate the goods wagons are generally heated. Serious losses often occur during the transport of seed potatoes from grower to merchant or if severe frosts set in after the pits have been opened. An ample supply of straw as insulating material may reduce these losses. Roots or tubers frozen in transit or in storage should be left undisturbed to thaw gradually, when it can be seen which have survived.

Those that have been killed can then be boiled, strained and used for fodder. In the case of shrubs and trees balanced manuring to ensure proper ripening and a thorough disease-control programme will go far to minimise frost damage. Damaged branches should be cut out to leave a clean, smooth surface to prevent their being used by fungi as avenues of attack.

DAMAGE CAUSED BY LIGHTINING

When lightning hits a tree it often only ploughs a furrow in the bark. Less frequently large pieces of bark are sheared off or the wood is splintered. Old trees with much decayed wood may even be set on fire. In a general way it may be said that tall trees are somewhat more likely to be struck by lightning than low ones, and in addition, those with rough bark like ash, oak, elm, willow, poplar and conifers tend to attract lightning more than trees with smooth bark, like sycamore, beech, birch and horse chestnut whose bark is more quickly and evenly wetted by rain, thus making them better conductors of electricity.

Lightning injury to growing potatoes has often been described. In a patch some three to ten yards across, where the lightning struck, the plants look as if they had been rolled or boiled. Round this is a zone of damaged haulm with flattened or split stems, and the pith may be killed, though the leaves still remain green.

When the injury occurs late in the season the tubers are also damaged. One end is often shattered and there are depressed blackened patches of flesh over which the skin may be still intact or broken later by shrinkage. These are apparently " burns " due to the passage of the electric current. Both on the haulm and in the tubers secondary attack by *Fusarium* and other fungi may occur.

SUN LIGHT

Light is essential to the life of green plants. Only in very exceptional cases, such as sycamore cotyledons, do tissues turn green in the absence of light. Moreover light is the source of energy whereby the green colouring matter (chlorophyll) is enabled to transform carbon dioxide and water into sugars and starch. The quality (colour) and intensity of light affect the growth of leaves and flowers and the process of transpiration.

Moreover, the daily duration of light is of great importance in determining whether a plant will flower or not (photo-periodism). Some plants flower only during long days; others, which normally flower only in the short days of autumn, can be induced to do so in summer if they are artificially darkened for the requisite number of hours a day.

IMPROPER SUN LIGHT

Lack of light generally induces etiolation (blanching). In extreme cases there is a yellow colour, long fragile stalk and small leaves; weak light induces straggling soft growth with few leaves, light green in colour.

Too dense a seeding in seedbeds and seed-boxes has a similar effect in producing weak, semi-etiolated plants particularly liable to damping off.

Injuries due to excessive sunlight are seldom seen and in practice it is difficult to differentiate between injury due to too intense a light and that due to the accompanying heat. Sun scald of greenhouse tomatoes in bright sunny weather is almost certainly due to the latter.

Too intense a light is, however, known to have a destructive effect on chlorophyll and it may be for this reason that plants suddenly moved into full sunlight make red pigment (anthocyanin) in their leaves. Such red colouring is regarded as abnormal in cacti and begonias. Aspidistras, camellias, *Cissits ant-arctica*, etc., do not flourish in strong light as their chlorophyll is gradually destroyed by it.

Tropical species of *Selaginella* with their delicate leaves cannot endure full sunlight, and many ferns require moderate shade. Camellias in strong light often develop yellow spots on their leaves.

MECHANICAL DAMAGE

Some wounds are caused intentionally, as during pruning of fruit trees and in lopping undesirable branches. All such wounds offer possible courts of entry to parasitic fungi and the larger ones should always be given a protective covering. Where branches have been torn off by wind the wound should be cut back to a smooth surface and covered in the same way. Accumulation of snow may break down many branches especially of evergreens and conifers.

Occasionally when light rain falls through a cold atmosphere it becomes supercooled and freezes immediately it touches a solid object. In such circumstances even small twigs may be surrounded by a layer of ice four inches thick.

DAMAGE BY HEAT

It is often difficult to distinguish heat damage from that due to too intense a light or to drought. During winter temperatures may be injurious to glasshouse plants which would be quite harmless with the brighter light of summer. A day's bright sunshine after a period of dull weather may scorch the leaves of plants under glass.

Shedding of flower buds and even of leaves may be a reaction to excessive heat, especially in azaleas and camellias. On outdoor plants, especially trees and bushes, hot dry weather is associated with high red-spider infestation. Pear leaves and apple and gooseberry fruits are all liable to sun-scorch. Sometimes raspberries may be sun-scalded and then have a grey hue which leads people to confuse them with mouldy berries. Most green leaves and stems are killed at temperatures of 100-120° F. though some succulents can endure rather more.

On hot summer days such high temperatures may occur near the surface of soil and rock. Some plants, such as flax and coniferous seedlings, are then liable to develop heat cankers at or near the soil level. In countries with a Continental climate, even trees may be thus injured near the base of their trunks. In such regions it is advisable to grow apple trees with very short trunks so that the branches always afford shade to the base of the trunk. When tree bark is marked by sun scald it is generally due to too rapid thinning of timber or the felling of trees that gave shade.

Sun scald of bark is not usually to be ascribed to the heat of the sun alone but to a combination of spring frosts and bright sunshine. Local scalding of leaves in glasshouse plants is sometimes due to an air bubble in the glass which acts as a lens concentrating the sun's rays on a single point. As the sun moves round this focus moves across the leaf leaving a seared line across it. Heating during storage is often blamed for decay in vegetables and potatoes but often unjustly.

Cool storage conditions are, however, desirable since vital processes are slowed down at low temperatures and the reserve food in tubers and similar storage organs is less rapidly used up. Evapouration is also slower at low temperatures and parasitic fungi and bacteria are usually less active. Storage at high temperatures, on the other hand, leads to premature shrivelling and sprouting.

Chapter 13

Plant Pathology

Having considered pathological conditions set up by factors of the environment and by viruses, fungi. These are all lowly forms of life. In organization they may be simple, but in many of them the details of their life-history are complex. They show an extraordinary variety of forms, and differ very much in physiology. One characteristic they all have in common is that none contain chlorophyll.

They may be green in colour, but the pigment is never capable of acting in photosynthesis. It must not be thought that all fungi, bacteria, and the other classes of these lowly plant organisms are pathogenic (disease-causing).Many are saprophytes, and do a useful service in breaking down dead remains of plants and animals.

Others are useful in that they "fix" gaseous nitrogen for other plants to use. Many are employed by man in commercial processes and form useful compounds for his use, yeasts in baking, brewing, etc. Finally, many, like the common mushroom, are edible and form a source of food. In agriculture, however, it is the pathogens attacking crops which attract the most attention.

All the organisms involved are, as has been said, devoid of chlorophyll, and because they attack the living plant are classed as parasites. Some of these parasites, after having killed the host, live on the corpse and are then saprophytes. All differ from a virus in that they are visible at some stage of their life either to the unaided eye or under some degree of magnification.

Perhaps the simplest kinds structurally are the bacteria in which the individual is a single microscopic cell. The cell varies in shape from spherical to rod-like. It may be supplied with little threads of protoplasm which thrash about like whips and are called flagella.

Each cell is simple as cells go, for it consists of a wall, protoplasm, and a nucleus so elementary in character as to be hardly worthy of the name. Reproduction in these unicellular plants is of two kinds. On the one hand, when conditions of food supply, temperature, etc., are suitable, each individual, by a process of simple fission, divides into two. This is quite devoid of any sexuality and is ajialagous to propagation in the higher plants. Division follows division quickly and multiplication is very rapid. On the other hand, when conditions become adverse, many forms of bacterium form a resistant structure.

When this occurs, the protoplasm of the cell shrinks away from the thin cell-wall. On the surface of the shrunken protoplast a very hard wall develops. Inside this hard resistant coat the protoplasm is protected from extremes of heat, destructive chemicals, and other agents which would have destroyed the vegetative stage. The resistant stage is called the resting spore. The bacteria are important in almost every aspect of agriculture; in the soil, the dairy, the cheese room; and as causes of disease in plants and animals. The fungi are very varied in structural organization, but are always rather more complex than the bacteria.

EFFECT OF ENVIRONMENT

When the disabilities of a plant more truly pathological in character are considered, three main classes of causative agent may be recognized.

These are:

1. Factors of the environment;
2. Viruses
3. Fungi and bacteria

The symptoms produced by different members of these three classes may be very similar in appearance, and considerable experience may be necessary to decide what is the cause of any particular trouble. To take quite a simple example, the brown necrotic patches produced in the soft leaves of a leek by a sharp hailstorm have been confused with lesions resulting from fungal attack or punctures made by an insect.

The symptoms induced in a plant by virus are easily confused with symptoms produced by deficiency of some nutrient in the soil. Frost damage in a potato crop may simulate very closely the effect of virus. Experience in differential diagnosis is required.

TEMPERATURE EFFECT

Extremes of temperature may cause pathological appearances. In hot climates "heat strain" affects many plants, and the rate of their various functions falls. This effect, attributed to the temperature, is often confused with that produced by over-intensity of light. Such a crop as coffee in many situations has to be protected by shade-trees interspersed throughout the growing area, while in glasshouse practice a coating of some form of whitewash is often applied to the glass in order to reduce the intensity of the summer sun.

The usual symptoms of over-intense heat or light are "bronzing" of the leaves, a peculiar metallic appearance with brownish-purple colouration.

EFFECT OF LOW TEMPERATURE

In agricultural practice in temperate regions it is lower rather than higher temperatures that usually cause trouble. A simple way in which this may happen occurs in winter when the soil-water is immobilized by being frozen to ice, though during the day-time at least the air conditions permit of rapid transpiration. Even leafless branches can lose a considerable amount of water through lenticels under these conditions.

Thus all the symptoms of drought may appear during a spell of freezing weather. Often it is shallow-rooted plants which suffer most. It has been shown that a plant with absorbing roots down in the not-frozen, deeper levels will continue to take in water satisfactorily. This water then moves through the upper roots unimpeded. Just as running water in a stream or pipe does not easily solidify, so the transpiration stream in the vasa passes up through the levels of the soil which are ice-bound.

Heaving

Mechanical effects of frost result in the plants "heaving." The ice expanding in the frozen earth presses the plant roots upwards into the air. When the thaw comes the soil volume returns to normal, and the upper part of the root is Jeft partially exposed.

The remainder of the root system still below the surface occupies tubular spaces in the soil. If drying winds follow the thaw the whole plant may die by drying out of the roots. Rosette plants with a conical tap-root, *e.g.* red clover, suffer most in this way.

A coating on the surface of the leaves and stem may impede gas exchange and cause asphyxiation, while the sheer weight of ice and snow accumulating on the shoot may cause mechanical breaking of the branches.

Frost Damage

"Winter-killing" of plants in the narrower sense of the term is due to action inside the tissues. At one time the death of cells in freezing weather was thought to follow from the expansion of ice crystals formed in the vacuoles bursting the cell-walls and protoplasts.

When the drop in temperature is sudden and goes to well below freezing point, damage may be due to formation of crystals inside the cells, but this is a rare occurrence in nature.

When the onset of freezing is comparatively gradual, as is usual under natural conditions, the ice crystals within the plant occur not in the cells but in the intercellular spaces.

The greater part of the water required to form these crystals must have been drawn from the vacuoles and protoplasts of the cells. Simple dehydration caused in this way may be the reason for the lethal effect of frost. Certainly the physico-chemical nature of the protoplasm is altered by freezing.

Frost Hardiness

The chief interest in these problems focuses on the fact that some plants can endure freezing temperatures and suffer no harm, while others cannot. lants can become "educated" to withstand low temperatures.

If the drop from moderate temperatures is accomplished gradually the plant becomes "hardened." In a plant treated in this way the dry matter content rises, as does the proportion of sugars to polysaxxharides. These results are probably associated with two different mechanisms. The first of these is that as temperatures fall, the rate of respiration falls faster than the rate of photosynthesis.

From this a larger "net credit balance" of dry matter accrues to the plant. Secondly, at lower temperatures the equilibrium point of the sugar-starch reaction swings over towards the sugar side. The net effect of both these physiological responses is to increase the sugar content of the tissues as a whole, and so strengthen their ability to hold water through an osmotic mechanism.

Along with these changes in the amount and nature of the dry matter there is a parallel development of the water-binding power of the colloidal material in the plant. The result of subjecting a plant to gradually falling temperatures is to strengthen its total power to hold water against any force tending to abstract it, and this confers an increased ability to withstand frost. Some species, varieties, or strains of plant under hardening conditions attain a greater degree of winter hardiness than do others. This is due to the fact that plants which have persisted for many generations in a cold climate have become harmonized to conditions there.

Plants whose ancestors have lived always under a more genial climate and have not been subjected to the same selective influences are not so adjusted. It is by natural selection over long periods of time that hardy and non-hardy strains have evolved. The reaction of both strains to hardening conditions is the same, though they differ in the extent to which the process can go.

The addition of soluble salts to the soil may increase the winter-hardiness of plants growing in it. This is probably due to an increase in the solute content of the cell sap increasing the water-holding power of the cells. It is to be noted, however, that over-supply of nitrogen to a plant renders it softer rather than hardier. In the plant different parts resist cold differently. On the same plant a young leaf is hardier than an older one. The leaf bases are more hardy than the tips, while the crown of such a plant as wheat is the hardiest part of all.

Thus, when frozen, the tips of old leaves die first, then the bases of the old leaf and the tips of the young leaves go, and finally the heart of the plant succumbs. The killed portions turn brown. Partial killing leaves a green centre to the plant, with the remainder brown or burnt.

Damage by Thawing

Thawing may cause damage in plants which have withstood freezing. When the thaw comes quickly crystals which have formed in the intercellular spaces revert rapidly to water. The cells appear to have neither the ability nor the time to take this fluid back into the cell, with the result that waterlogging of the tissues and rotting follow.

"Blebs" or blisters full of fluid often appear in the leaf surface under these conditions. This is often seen in early potatoes frosted during the night. The morning sun beats on one side of the drills, and causes a rapid thaw on that side of the frozen plants.

The leaves of the sunny side thaw quickly and are killed. The leaves on the dull side thaw more slowly, time is given to the tissues to adjust, and so they may be less severely affected.

Effects of Hail

Hail, as has been said, may, by its hitting power, kill patches of tissue on soft succulent leaves. The whole plant may be beaten down flat by hail, or by heavy rainstorms or boisterous winds.

Tissues Burning

Drying winds often "burn" the tips and margins of leaves by excessive water-loss. This is especially so in young tissue where the cuticle is not yet sufficiently developed as to prevent the passage of water across the epidermis. Spray-laden winds blowing in off the sea may deposit common salt on leaves to such an extent that plasmolysis and death of the cells occurs.

Effect of Air Pollution

Near large towns noxious fumes, often of a sulphurous nature, may cause death or damage of leaf tissues. Sooty and tarry smokes in the air, when deposited on leaves, reduce their efficiency.

The stomata may be blocked by such solids, and definite damage or death may result. In the soil the water/air supply to the roots may be unsatisfactory, so that the living cells there do not obtain sufficient oxygen. The living roots die or are narcotized in these circumstances, and water intake is considerably affected.

SOIL IMPURITIES

The disabilities of the plant due to external factors so far discussed are usually readily recognized, and the cause is fairly obvious. An external factor not so readily recognized, and which produces symptoms more easily associated with other agents, is an unsuisupply of soluble substances in the soil. The unsuitability may be due to the presence of a substance definitely poisonous to the plant. Such a case rarely arises in practice.

The kind of state which commonly occurs is imbalance of the nutrient substances, which when in normal proportions are necessary for the health of the plant.

Improper Salt Effects

Imbalance may be due to simple deficiency of an essential element of either the macro or micro class'. On the per cent other hand, it may be due to an overplus of one such ingredient causing relative deficiency of another. Excess nitrogen may render an otherwise adequate supply of phosphorus or potash insufficient for the abnormal demands of the plant. Too much lime may accentuate a partial deficiency of boron.

There is an ideal balance between any one nutrient and the others. Apart from differences in the absolute amount required, the main difference between macro and micro elements lies in the fact that the range in quantity of supply which the plant will tolerate is greater in macro than in micro elements. The symptoms of relative over-supply or under-supply of each of the more important elements may be briefly summarized.

Nitrogen

When nitrogen is deficient, apical meristems become inactive and root and shoot elongation is arrested, lateral buds remain dormant, and there are few branches. Tillering in cereals is depressed.

The whole plant becomes pale yellowish-green, and later this colour intensifies so that leaves turn yellow, orange-red, or purple. Flower production is restricted. The plants generally are stunted and discoloured. Over-abundant nitrogen, on the other hand, produces a large, soft, succulent growth, dark green in colour, and often susceptible to insect and fungus attack.

Phosphorus

Phosphorus, like nitrogen, is required for protein formation, and when deficient results in the production of a stunted plant with few branch shoots and a very poorly developed root system.

The symptoms of these two deficiences (N and P) are very similar except that with phosphorus deficiency the leaf colour is a dull, bluish-green. In some cases the colour produced may resemble nitrogen starvation in being purple, but it is rarely yellowish or red. The leaf margins may show scorching in the absence of phosphorus.

Calcium

With short supply of this element the young leaves become considerably distorted, rolled, and curled-up with ragged edges. The root system is poorly developed and stumpy. In the cabbage and turnip class of plant the chlorophyll of the shoot is deficient or absent, and chlorosis is said to occur. The margins of the leaf may be quite white. Clovers and related plants are sensitive to calcium shortage.

Red clover, cannot take up soluble calcium from a medium of high hydrogen ion concentration. Legumes in lime-deficient soil show white-spotting chlorosis, and the petioles of the leaves bend or "kink" in the middle region, allowing the blade to droop or hang down.

With these plants the position in regard to this element is complicated, because the acidity-intensity of the soil (H.I.C.) affects nodule formation on the roots, and these are also concerned in the production of symptoms. Tuberization of the potato may be quite inhibited in a soil extremely deficient in lime.

Magnesium

Plants growing in a soil deficient in magnesium show chlorotic symptoms which commonly commence in the older leaves. This symptom is probably connected with the fact that magnesium is required for chlorophyll formation, and is actually the only metallic component of the molecule.

The symptoms of magnesium deficiency vary very considerably as between different kinds of plants. The members of the cabbage family seem to show the effects of shortage of magnesium before others. Cauliflower, broccoli, and kale are most susceptible. Chlorotic marbling or mottling of the leaves is a first symptom, followed later by the appearance of bright yellow, red, orange, and purple colouration.

In swedes magnesium deficiency is difficult to tell from phosphorus deficiency owing to the reddish-purple colour produced in both cases. In potato different varieties show different detailed symptoms of magnesium deficiency, but usually these are all chlorotic or necrotic in character. It has been stated that "blight" is more severe on potatoes growing on magnesium-deficien t soil. Tomatoes grown under glass are often given heavy dressings of potash. This brings on a marked early chlorosis typically symptomatic of deficiency of magnesium in this plant.

Potassium

The most obvious symptom of potassium deficiency is a scorch or burn at the leaf tip or leaf margin. The plants are stunted, and the leaf colour is a dull, greyish blue-green. The general appearance of lack of vitality in the plants is typical. Clovers and their allies show pronounced symptoms in potassium deficiency. In addition to marginal browning, clover leaves in the early stages of the trouble show a crescent-shaped brown arc with pronounced spotting.

Sulphur

Deficiency of this element is indicated by chlorosis and browning of the leaves; the buds die and plants are stunted. In many crops the general appearance is very similar to that following nitrogen deficiency, but in the case of lack of sulphur the roots have a peculiar brown colouration.

Iron

Iron deficiency is usually associated with high lime content in the soil, and hence the chlorosis which is typical of deficiency -of iron is often referred to as lime-induced chlorosis. The excess lime present affects the rate of iron intake. In cereals, the chlorosis is often accompanied by a dying of the leaf tips.

Manganese

In oats, "grey speck disease," often mistaken for the symptoms of fungus attack, is caused by lack of manganese. Greyish specks or streaks appear on the lower half of the leaves when the plant is four or six inches high. In addition to the grey specks, the leaf doubles over with a transverse fold. The roots develop poorly.

The leaves of wheat and barley plants subjected to manganese deficiency are faintly chlorotic and may be streaked with yellow. In crops such as kale, cabbage, and their allies, the leaf may be entirely chlorotic except for the veins, which remain green. In response to lack of manganese the condition known as "speckled yellows" appears in sugar beet. The seed of peas grown on manganese-deficient soil shows a characteristic brown spot on the morphologically upper face (inward facing surface) of the cotyledons. This condition is usually known as "marsh spot."

Boron

The symptoms produced by boron deficiency vary. Cereals seem to have a very low requirement for boron, and hence they rarely show well-defined symptoms. "Root" crops on boron-deficient soil show browning or "rotting" of the inner tissues of the "bulb." Cauliflowers are very susceptible to a poor supply of this element, and indicate a lack by failing to develop a "curd." If the deficiency is not extreme the curd turns brown.

Cabbages, curiously enough, seem to be unaffected. An over-supply of boron can be as bad as an under-supply. In the potato, over-supply of boron may prevent "germination" of the "seed," or it may inhibit development of the sprouts and their adventitious roots. The leaves of maize show white chlorotic tips when grown in a soil deficient in this metal. The condition set up is usually referred to as "wither tip."

Copper

Lack of this element in the soil seems to affect the herbage growing on it less than it does the grazing animal. "Falling-down" disease of cattle in West Australia is associated with deficiency of this metal in the soil; animals grazing on copper deficient herbage when driven fall down and die. The twigs of fruit trees growing on certain soils die back progressively from the tips; the leaves wither and die.

The "disease" is known, generally, as "die-back" or exanthema, and is readily cured by injection of copper salts into the root crown or spraying the leaves with soluble copper compounds. In cereals marginal chlorosis yellowing of the blade and withering of the leaf tips is associated with copper deficiency. This, because it often appears in the first crop on heath and peat soils when they are taken into cultivation, has been called reclamation disease.

MINERAL DEFICIENCY DIAGNOSIS

From these brief outlines it will be realised that the differential diagnosis of the different deficiencies by eye judgment alone may be very difficult. Other more accurate but more labourious methods are available, a number of which may be described. Small plots or areas in affected crop are marked off, and watered or sprayed with a weak solution of a salt supplying one of the metals suspected of being deficient.

The leaves should be thoroughly wetted. In twelve to fourteen days the plot receiving that element which is deficient in the soil will show a reduction of symptoms. The method is most successful if practised on a young crop.

Plant's Indicator

The use of indicator plants for testing soil has been suggested. On a soil suspected of manganese deficiency oats may be grown, and the typical grey speck will appear if the element is really in short supply. Marrow-stemmed kale, broccoli, and cauliflower all will show very characteristic symptoms when grown in soil deficient in nitrogen.

In agricultural practice, the troubles arising from an imbalance of nutrient supply usually arise in a commercial crop when there is little time for elaborate testing if the crop is to be saved and a normal harvest secured. If the trouble is noticed soon enough, that is, when the crop is young, manurial dressings of an appropriate mixture of salts may be applied.

Once the deficiency has been accurately diagnosed, the manuring programme for the future should be suitably modified. When the symptoms of deficiency diseases are compared with those due to the presence of a virus, it will be seen that differentiation in diagnosis between the two classes of cause may be as difficult as between two deficiencies.

VIRUS EFFECT

Distinction between a deficiency condition and a virus disease condition lies not so much in the actual symptoms exhibited by the affected plants but in the fact that, if the cause be a virus, the trouble may be transmitted from an unhealthy to a healthy plant. The methods adopted vary.

Some cases require the normal, healthy plant to be grafted on to the unhealthy one. In other cases simple inoculation of the healthy plant with sap from a virus-infected individual is effective. If the cause of ill-health is due to a deficiency of an element the healthy plant will remain healthy no matter what method is used. Virus is transmissible, deficiency is not. Deficiency is readily diagnosed and cured by the use of an appropriate salt solution or manurial treatment. Virus symptoms in general do not answer to treatment.

WHAT IS VIRUS

The plant viruses are best known from their effects on the plants which they attack; none of them has ever been cultivated outside the plant body. They are not visible in a microscope using visible light, for their structure is ultra-microscopic. Some viruses have been isolated from the host, and all have proved to be some form of nucleic acid; they are of the nature of nucleo-proteins.

These extracted "chemicals" may be stored outside the plant and retain their virulence for varying periods. The viruses then may be defined as pathogens, always parasitic, and composed of units ultra-microsocopic in size. Viruses being parasitic in nature, a plant which bears them may be called the host.

SYMPTOMS OF VIRUS ACTION

The symptoms of attack by virus are very variable, for the appearances produced depend not only on the nature of the virus but on the nature of the host. The commonest type of symptom is a deformation of the shoot, a crinkling and rolling of the leaves, more generally distributed over the plant than is seen in fungal attack.

These malformations are often accompanied by discolourations of the leaf and stem. The same virus may cause quite different symptoms in different plants, and indeed in different varieties of the same plant, while different viruses may be the cause of identical symptoms in one host. A plant may be fully infected with a virus and yet show no obvious symptoms. Such a plant is known as a "carrier," and it is fully capable of infecting a susceptible neighbour.

The fact that the virus is present in a "carrier" can only be demonstrated by transmission experiments. Symptoms of some viruses in certain hosts may not be visible externally, but various bodies may appear in the host tissues. These bodies may appear as crystals or simulate such lowly organisms as amoeba. The tissues of the plant may be altered. The chloroplasts may be damaged or reduced in number, or necrotic lesions appear in the phloem. Symptoms of first or primary infection with virus always show in tissue developed after infection.

The trouble seems to spread through the host from a centre at the point of entry to areas where growth is active, and all tissues developed thereafter show the symptoms, while older organs remain normal. When a propagant such as a potato tuber contains the disease the whole plant developed from it shows symptoms; this is not primary infection.

TRANSMISSION

Once inside a congenial host-plant the virus increases and moves about the plant body very much as food metabolites do. The sap is now infective and may be carried from one plant to another.

A diseased plant rubbing against a healthy one with very slight bruising of each may serve. The hands of a worker touching a diseased plant can carry the pathogen to a healthy one touched subsequently. A plant parasite such as dodder acting as an organic link between two plants can provide a channel for the virus.

Vector

It is an insect, however, which is most commonly the carrier of infection. If an appropriate insect sucks fluid from the tissues of a diseased plant and then proceeds to a healthy individual, it regurgitates into the second, and so infection is spread.

The more mobile the insect the greater the rate of spread. It should be remembered that small insects which are themselves but weakly mobile, may be carried great distances by wind. Any agent, insect or not, capable of transmitting the virus is called a *vector*. A quite sparse infestation of an insect, not serious as a pest in itself, may be amply sufficient to transmit virus quickly from plant to plant.

In some cases at least the insect vector does not act simply as a transmitter. Evidence is accumulating that while the virus is in the insect it undergoes some change or multiplies. Thus, while in many cases an insect which has fed on a diseased plant is immediately capable of infecting a healthy one, this is not necessarily so; a certain period of time may have to elapse between the two incidents for transmission to be effective.

VIRUS CONTROL

Control of virus disease takes two lines: prevention and cure. Prevention depends for success on commencing with a healthy stock of plants. This attended to, any source of virus infection existing in the neighbourhood should be eliminated. Weeds often carry virus. If complete elimination of infected plants in the vicinity is not possible, the vector should be eliminated or its movement prevented.

It is doubtful if the embryo in a seed can carry virus, though it is believed that "bean mosaic" is transmitted in this way, and quite a number of cases are known where the coverings of the "seed" carry disease. The very young seedlings developed from these seeds are quite healthy, though a suspicion exists that infection of the plantlet takes place after germination, the virus passing from the covering to the emerging axis.

Even a few unhealthy plants produced in a crop in this way are dangerous, for they act as sources for the primary infection of the healthy members. A propagant taken from a diseased plant is almost certain to be infected. Potato tubers, sugar-cane sets, strawberry runners, or raspberry off-sets are all highly dangerous if derived from diseased crops.

A partial safeguard lies in obtaining such pro-pagants only from a healthy crop growing in an area where the appropriate vectors are scarce. This does not provide absolute protection, for a crop inspected early and certified as clean may have become infected late in the season.

In this case the plants would not have had enough time to develop observable symptoms when inspected. This may be guarded against by growing a representative sample of the propagant in a glasshouse well before the main bulk is due to be planted in the field. This is done in the production of "elite" stocks. In practice, for commercial purposes, the purchase of seed from a crop proved free of disease by inspection at various times in the growing season should be adequate.

Virus Infection

Weeds in an area often maintain a "reservoir" of disease ready to act as a source for the primary infection of a crop. Not only those weeds botanically related to the crop should be regarded as suspect, but any weed should be mistrusted. One and the same virus may live and develop quite well in two very dissimilar hosts—*e.g.* tomato and chrysanthemum, or potato and aster.

So too, as has already been said, some plants, weeds or otherwise, tolerate the presence of a virus but show no symptoms. These are quite capable of supplying material for the infection of a susceptible neighbour.

So much for elimination of sources of primary infection; the alternative is prevention of transmission. This is usually difficult. The most obvious step is to clean up the breeding ground of such common vectors as greenfly, capsid bug, etc. Where practicable, sowing the crop earlier or later than usual may "dodge" the incidence of the insect. Valuable horticultural subjects, or elite stocks of more humble plants, may be protected from the vector when grown under gauze screens, or grown in remote areas which are free from infective material or vectors.

In horticultural practice the human hand, pruning knives, seed boxes, or roots and plant debris in die soil can act as vectors. Appropriate forms of control, as by systematic washing of the hands, disinfection of knife blades, heat treatment of boxes and soil, should be introduced.

Cure of Virus Disease

Curative measures applied to a virus diseased plant are not likely to prove feasible in general agriculture. Various chemicals have been suggested as active against virus, but their use is hardly practicable. Heating infected plants, especially when they are in a dormant condition, promises well as a method of cure in a limited number of cases.

A number of quite serious viruses are reduced in virulence or "killed" by a temperature which does not harm the plant. In other cases the virus remains unaffected by temperatures which kill the plant. Stocks or scions of fruit-trees, and in other crops propagants such as tubers, bulbs, or corms may be treated in this way.

Virus Immunity

The breeding of varieties immune to all kinds and strains of kinds of virus is an ideal which, probably, never will be attained, though resistance to.some is possible. Two forms of immunity can be recognized: conferred immunity, and inherited or genetical immunity.

The outstanding success so far in breeding for virus immunity has been achieved in Java, where the P.O.J. varieties of sugarcane are resistant to "sugar-cane mosaic." Such plants become infected but show no symptoms and produce a full yield. They behave as carriers, however, and can transmit the disease to susceptible varieties. Varieties which tolerate the presence of a virus in this way may be a menace to the agriculture of an area.

Unless they and they alone are grown in any one isolated situation, they act as a reservoir of infective material, and a supply of the pathogen is available to all susceptible varieties growing near by. The matter does not end here, for, as has been shown, one and the same virus may infect a number of hosts otherwise unrelated.

Hence, the presence of such a non-susceptible variety of one crop may result in disease attacking a different crop cultivated in the same area. Paradoxically, it may be that the best control may come from the production and cultivation of highly susceptible varieties, and this for either of two reasons. If a highly susceptible variety is grown the "seed" crop can be "rogued" very efficiently. Any member becoming infected shows the symptoms quickly and clearly, and so is soon eliminated. Some varieties are so susceptible to virus that as soon as primary infection takes place the tissues round the site die.

In these cases death of the tissues results in either the death of the pathogen or inability of the virus to diffuse out to other parts. In these hypersensitive forms primary infection results only in local lesions, and the trouble does not affect the plant generally. In addition to these aspects of inherited immunity and/or susceptibility there is the question of conferred immunity. This appears usually in cases where one virus negatives the effect of another when both are present in the same host.

The whole question of immunity is overshadowed by the knowledge that a virus may "mutate" freely, and so change easily in almost all its characters, including its powers of infectivity and the symptoms produced in a given host. These complications of the virus position have not been simplified by the difficulties of naming the various kinds and their strains.

The method up-to-date has been to give the virus a name indicating its host and the more obvious symptom it produces. The virus which causes yellowish-white or greenish transparent patches to appear on the leaves of potato has been called "potato mosaic." This method was satisfactory until it was found that not one but many viruses could each produce "mosaic" in the leaf of a potato.

Furthermore, the potato viruses are not confined to potato or even to its relatives in the genus *Solanum*. They infect other and quite unrelated hosts, and produce in them symptoms other than mosaic. Classifying viruses by the symptoms they produce or the hosts they infect is not satisfactory. Classification on the basis of their usual vectors breaks down too. The aphid *Myzus persicae* alone transmits at least twenty-six different viruses.

A recommendation made to the International Botanical Conference suggested a nomenclature based on the name of the host followed by the word virus, then a number. Strains of the virus would then be indicated by a letter; thus, cucumber virus. The sub-strains of this indicated by a small letter, *e.g.* "a." This semi-official recommendation has not yet been adopted generally, and in practice the original method of naming by reference to the most usual host and the most usual symptoms still holds.

FUNGI EFFECT

A member of the true fungi always shows a definite structural organization. The unit of the vegetative body is the *hypha,* a long, thin, hair-like filament consisting of a wall lined with protoplasm.

The hypha in some members is divided by cross walls, when it is said to be septate, or there may be no septa present. The a-septate hypha has many nuclei and is said to be coenocytic. In the septate forms the individual segments may contain one, two, or many nuclei. These latter types are also described as coenocytic.

When many hyphae go to make up one plant they are referred to collectively as the *mycelium.* The mycelium never shows any differentiation into root, stem, or leaf, but there may be regions showing some degree of specialization for function. Part of the mycelium may be purely vegetative, while another portion is specialized for spore production. This is clearly seen in the familiar mushroom, where there is a white mycelium ramifying through the soil, collecting food and water. Above ground there appears the umbrella-like, spore-producing organ.

The underground vegetative portion sometimes forms a white cord-like similitude of a root, this is called a rhizomorph. It is to be remembered that the whole of the mushroom plant, in common with all other fungi, is constructed of hyphae, loose in one part and compacted for a special purpose in another. The compacted portion of these, when examined microscopically, simulates a cellular tissue and is known as pseudo-parenchyma.

PREPRODUCTION OF FUNGI

The increase and spread of these plants is accomplished by means of the vegetative mycelium, asexual spores and sexual spores. The mycelium, when broken up, functions as a propagant just as parts of the body may do in higher plants. This is taken advantage of in mushroom growing, when a mass of mycelium in a suimedium (called "spawn ") is used to propagate the crop.

Asexual spores may be looked upon simply as small portions budded off from the mycelium and are therefore another but more specialized method of vegetative propagation. The sexual spores may be regarded as the fungal parallel of the seeds of higher plants, in which case the vegetative mycelium and its non-sexual spores may be regarded as the functional parallels of rhizomes and corms in propagation.

The disease-causing fungi, being parasites, are always faced with difficulty in completing their life-cycle, for in order to do so a suihost must be found. Reproductive units must be very freely produced in order to ensure that some at least are successful. The spore, which is the unit in reproduction, is almost invariably one-celled, not commonly two-celled, and is produced in enormous numbers. Different kinds of spores may be produced at different points in the life-cycle, but the total number is always very, very high.

Fungal Spore

There are, as we have seen, two general classes of fungal spore. There are those which form by mere budding off from a hyphal tip and are vegetative spores. These are typically thin-walled and not resistant to adverse conditions. They are used for the rapid multiplication of the plant during periods favourable to the vegetative phase.

These may be called summer spores, asexual spores, etc. On the other hand, there are the spores produced as a sequel to or as the result of a sexual act or fusion of gametes. The sexual mechanism in the fungi is often obscure and complicated, and in many cases is not well understood. Spores produced sexually are often capable of remaining dormant for long periods of time, or resisting conditions adverse to the vegetative state.

These may be called sexual, resting, or resistant spores. The terms often refer to the method of their production, to a feature of their structure and appearance, or to the nature of the "fruiting" body in which they are produced. When the spores are simply budded off from the tips of hyphae they are called *conidia* or conidial spores. Among the sexually produced spores, two broad classes may be remembered. Those produced by isogamy (fusion of morphologically identical gametes) are zygospores; those produced by heterogamy (fusion of dissimilar gametes) are referred to as oospores. Sometimes a special hypha or a group of special hyphae is formed, in which a sexual act takes place.

From each hyphal end four little filaments are produced. Each bears a small spore or sporidium like a bust on a pedestal. The hyphal end is called a basidium and the sporidia produced are called basidio-spores. When the basidia are massed in a special layer this is called a hymenium. Another fructification is the ascocarp, which has a layer of sexually formed flask-shaped cells, each called an ascus (plural, asci). Each ascus contains up to eight peculiar spores, the ascospores.

The important points in connection with the increase and spread of disease-causing fungi are that reproduction is predominantly by spores; that they are produced in enormous numbers; and that they must reach a suihost if they are to function. Some fungi are strict in their requirements as to the kind of host they can infect; that is, they are specific to one host. Others can infect a wide range of host plants; they are nonspecific. A few, such as some rusts to be described later, must infect two hosts alternately, passing by special spores from one to the other in each complete life-cycle. These last are said to be heteroecious.

FUNGI CONTROL

Protection of the crop plant against disease follows two main principles, prevention and cure. The whole study of protection will be dealt with here, first genera.

Fungal Parasites

Probably no method of protecting a crop against fungal disease is so attractive as the production and growth of an immune variety of host. If a variety completely immune to all diseases could be evolved in each crop, the practice of agriculture would be considerably simplified.

This is hardly possible, but varieties showing heriresistance to a specific disease are not only possible but are now well known. "Conferred immunity" acquired by a plant and lasting for the life of that generation only is also known. When considering an immunity of a crop plant to a disease three main points must be kept in mind. In the first place, immunity to one disease does not confer immunity to another. Resistance to each of the diseases to which the plant is heir must each be bred for or developed separately.

Hence, in writing of an immune variety, what is referred to is a variety which will not readily contract some specified complaint. Secondly, complete and total immunity to a single disease in an otherwise susceptible crop probably does not exist.A rather more accurate view is that some individuals or varieties are more resistant or less susceptible to a particular disease than are others.

If, however, the pressure of infection on these resistant forms is increased past a critical point, or if environmental conditions are highly favourable to the parasite or adverse to the host, then the fungus may gain the upper hand. Thirdly, the fact that in disease there are two components, the host and the parasite, must never be lost sight of.

Just as resistant forms of the host may develop due to genetical alterations in it, so too changes of a hericharacter may take place in the parasite which render it more virulent and capable of breaking down a previously existing immunity. Many disease organisms have developed a number of strains or biological races, each showing different degrees of virulence but all morphologically alike.

So a particular crop plant variety may be immune to one strain of a fungus but be quite susceptible to another strain. Very often the strains of different virulence, or different ability to attack, inhabit different districts or countries. Thus a crop plant variety immune to a definite disease in one geographical area may become infected by that disease when transferred to another area.

So, too, the inadvertent transportation of a strain of fungus from its "home" territory to a new area may cause disease of the crops long resident there and immune to the local form of the pathogen. A case of this was seen when the wheat variety "Bobs" was taken from Australia to India and South Africa. Bred in Australia, "Bobs" had proved to be immune To the black rust there. In India and South Africa the immunity broke down completely.

Either the black rust fungus of India and South Africa is of a different biological race from that of Australia, or growth in the new environment had so altered the wheat that the resistance mechanism failed.

Index